NEWS

NASA

NATIONAL AERONAUTICS AND SPACE ADMINISTRATION
WASHINGTON, D.C. 20546

TELS. WO 2-4155
WO 3-6925

FOR RELEASE: SUNDAY
December 15, 1968

RELEASE NO: 68-208

PROJECT: APOLLO 8

I0040866

contents

-0-

12/6/68

Cover: Greetings From Apollo 8 Christmas Eve, Dec. 24, 1968: Mission commander Frank Borman, command module pilot James Lovell and lunar module pilot William Anders gave the world a holiday greeting from lunar orbit, as the Apollo 8 crew became the first humans to leave Earth's gravity and reach the moon. In the image, the Apollo 8 crew stands in the doorway of the recovery helicopter after arriving aboard the carrier U.S.S. Yorktown. In left foreground is Commander Borman. Behind him is Lovell and on the right is Anders. Apollo 8 splashed down at 10:51 a.m. EST, Dec. 27, 1968, in the central Pacific Ocean, approximately 1,000 miles south-southwest of Hawaii.

Image Credit: NASA

Published by Books Express Publishing
Copyright © Books Express, 2012
ISBN 978-1-78039-857-0

Books Express publications are available from all good retail and online booksellers. For publishing proposals and direct ordering please contact us at: info@books-express.com

FOR RELEASE: SUNDAY
December 15, 1968

RELEASE NO: 68-208

FIRST MANNED LUNAR ORBIT MISSION

The United States has scheduled its first mission designed to orbit men around the Moon for launch Dec. 21 at 7:51 a.m. EST from the National Aeronautics and Space Administration's John F. Kennedy Space Center, Florida.

The mission, designated Apollo 8, will be the second manned flight in the Apollo program and the first manned flight on the Saturn V rocket, the United States' largest launch vehicle.

Crewmen for Apollo 8 are Spacecraft Commander Frank Borman, Command Module Pilot James A. Lovell, Jr. and Lunar Module Pilot William A. Anders. Backup crew is Commander Neil A. Armstrong, Command Module Pilot Edwin E. Aldrin, Jr. and Lunar Module Pilot Fred W. Haise, Jr.

-more-

12/6/68

Apollo 8 is an open-ended mission with the objective of proving the capability of the Apollo command and service modules and the crew to operate at lunar distances. A lunar module will not be carried on Apollo 8 but Lunar Test Article (LTA-B) which is equivalent in weight to a lunar module will be carried as ballast.

The mission will be carried out on a step-by-step "commit point" basis. This means that decisions whether to continue the mission or to return to Earth or to change to an alternate mission will be made before each major maneuver based on the status of the spacecraft systems and crew.

A full duration lunar orbit mission would include 10 orbits around the Moon. Earth landing would take place some 147 hours after launch at 10:51 a.m. EST, Dec. 27.

Earlier developmental Apollo Earth-orbital manned and unmanned flights have qualified all the spacecraft systems --including the command module heat shield at lunar return speeds--and the Apollo 7 ten-day failure-free mission in October demonstrated that the spacecraft can operate for the lunar-mission duration.

-more-

Apollo 8 will gather data to be used in early development of training, ground simulation and crew inflight procedures for later lunar orbit and lunar landing missions.

The Dec. 21 launch date is at the beginning of the December launch window for lunar flights. These windows hinge upon the Moon's position and lunar surface lighting conditions at the time the spacecraft arrives at the Moon and upon launch and recovery area conditions. The December window closes Dec. 27. The next comparable window opens Jan. 18 and closes Jan. 24.

The mission will be launched from Complex 39A at the Kennedy Space Center on an azimuth varying from 72 to 108 degrees depending on the launch date and time of day of the launch. The first opportunity calls for liftoff at 7:51 a.m. EST Dec. 21 on an azimuth of 72 degrees. Launch of Apollo 8 will mark the first manned use of the Moonport.

The Saturn V launch vehicle with the Apollo spacecraft on top stands 363 feet tall. The five first-stage engines of Saturn V develop a combined thrust of 7,500,000 pounds at liftoff. At igintion the space vehicle weighs 6,218,558 pounds.

Apollo 8 will be inserted into a 103 nautical mile (119 statute miles, 191 kilometers) Earth orbit.

During the second or third Earth orbit, the Saturn V third-stage engine will restart to place the space vehicle on a path to the Moon. The command and service modules will separate from the third stage and begin the translunar coast period of about 66 hours. A lunar orbit insertion burn with the spacecraft service propulsion engine will place the space-craft into a 60 x 170 nm (69 x 196 sm, 111 x 314.8 km) elliptical lunar orbit which later will be circularized at 60 nm (69 sm, 111 km).

The translunar injection burn of the third stage will place the spacecraft on a free-return trajectory, so that if for some reason no further maneuvers are made, Apollo 8 would sweep around the Moon and make a direct entry into the Earth's atmosphere at about 136 hours after liftoff and land in the Atlantic off the west coast of Africa. During the free-return trajectory, corrections may be made using the spacecraft Reaction Control System.

Ten orbits will be made around the Moon while the crew conducts navigation and photography investigations. A trans-earth injection burn with the service propulsion engine will bring the spacecraft back to Earth with a direct atmospheric entry in the mid-Pacific about 147 hours after a Dec. 21 launch. Missions beginning later in the window would be of longer duration.

Several alternate mission plans are available if for some reason the basic lunar orbit cannot be flown. The alternates range from ten days in low Earth orbit, a high-ellipse orbit, to a circumlunar flight with direct Earth entry.

As Apollo 8 leaves Earth orbit and starts translunar coast, the Manned Space Flight Network for the first time will be called upon to track spacecraft position and to relay two-way communications, television and telemetry in a manned spaceflight to lunar distance.

Except for about 45 minutes of every two-hour lunar orbit, Apollo 8 will be "in view" of at least one of three 85-foot deep-space tracking antennas at Canberra, Australia, Madrid, Spain, and Goldstone, California.

Speculation arising from unmanned Lunar Orbiter missions was that mass concentrations below the lunar surface caused "wobbles" in the spacecraft orbit. In Apollo 8 the ground network coupled with onboard navigational techniques will sharpen the accuracy of lunar orbit determination for future lunar missions.

Another facet of communicating with a manned spacecraft at lunar distance will be the use for the first time of the Apollo high-gain antenna--a four-dish unified S-band antenna that swings out from the service module after separation from the third stage.

The high-gain antenna relays onboard television and
high bit-rate telemetry data, but should it become inopera-
tive, the command module S-band omni antennas can relay voice
communications, low bit-rate telemetry and spacecraft commands
from the ground.

Apollo 8 will gather data on techniques for stabilizing
spacecraft temperatures in deep-space operations by investi-
gating the effects of rolling the spacecraft at a slow, fixed
rate about its three axes to achieve thermal balance. The
Apollo 8 mission will be the first opportunity for in-depth
testing of these techniques in long periods of sunlight away
from the reflective influence of the Earth.

Any solar flares occurring during the mission will be
monitored by Solar Particle Alert Network (SPAN) stations
around the world. Solar radiation and radiation in the Van
Allen belt around the Earth present no hazard to the crew of
Apollo 8 in the thick-skinned command module. The anticipated
dosages are less than one rad per man, well below that of a
thorough chest X-ray series.

Although Apollo 8's entry will be the first from a
lunar flight, it will not be the first command module entry
at lunar-return velocity.

The unmanned Apollo 4 mission in November 1967 provided a strenuous test of the spacecraft heatshield when the command module was driven back into the atmosphere from a 9,769 nautical mile apogee at 36,545 feet-per-second. By comparison, Apollo 8 entry velocity is expected to be 36,219 feet-per-second. Heatshield maximum char depth on Apollo 4 was three-quarters of an inch, and heat loads were measured at 620 BTUs per square foot per second as compared to the 480 BTUs anticipated in a lunar-return entry.

Apollo 8 entry will be flown with a nominal entry range of 1,350 nautical miles in either the primary or backup control modes. Adverse weather in the primary recovery area can be avoided by a service propulsion system burn prior to one day before entry to shift the landing point. Less than one day out, the landing point can be shifted to avoid bad weather by using the spacecraft's 2,500 mile entry ranging capability.

The crew will wear the inflight coveralls during entry-- pressure suits having been doffed and stowed since one hour after translunar injection. Experience in Apollo 7, when the crew flew the entry phase without pressure suit, helmets or gloves, prompted the decision not to wear suits once the spacecraft's pressure integrity was determined.

The decision to fly Apollo as a lunar orbit mission was made after thorough evaluation of spacecraft performance in the ten-day Earth-orbital Apollo 7 mission in October and an assessment of risk factors involved in a lunar orbit mission. These risks are the total dependency upon the service propulsion engine for leaving lunar orbit and an Earth-return time as long as three days compared to one-half to three hours in Earth orbit.

Evaluated along with the risks of a lunar orbit mission was the value of the flight in furthering the Apollo program toward a manned lunar landing before the end of 1969. Principal gains from Apollo 8 will be experience in deep space navigation, communications and tracking, greater knowledge of spacecraft thermal response to deep space, and crew operational experience--all directly applicable to lunar landing missions.

As many as seven live television transmissions may be made from Apollo 8 as it is on its path to the Moon, in orbit about the Moon and on the way back to Earth. The television signals will be received at ground stations and transmitted to the NASA Mission Control Center in Houston where they will be released live to commercial networks.

However, because of the great distances involved and the relatively low transmission power of the signals from the spacecraft to ground, the TV pictures are not expected to be of as high quality as the conventional commercial broadcast pictures.

Apollo 8 also will carry still and motion picture cameras and a variety of different films for black and white and color photography of the Moon and other items of interest. These include photographs of an Apollo landing site under lighting conditions similar to those during a lunar landing mission.

The Apollo 8 Saturn V launch vehicle is different from the two unmanned rockets that have preceded it in the following major aspects:

The uprated J-2 engine capable of reaching a thrust of 230,000 pounds is being flown for the first time, on the third (S-IVB) stage.

The new helium prevalve cavity pressurization system will be flying on the first stage (S-IC) for the first time. In this system, four liquid oxygen prevalves have cavities filled with helium to create accumulators "shock absorbers" that will damp out the "pogo" effect.

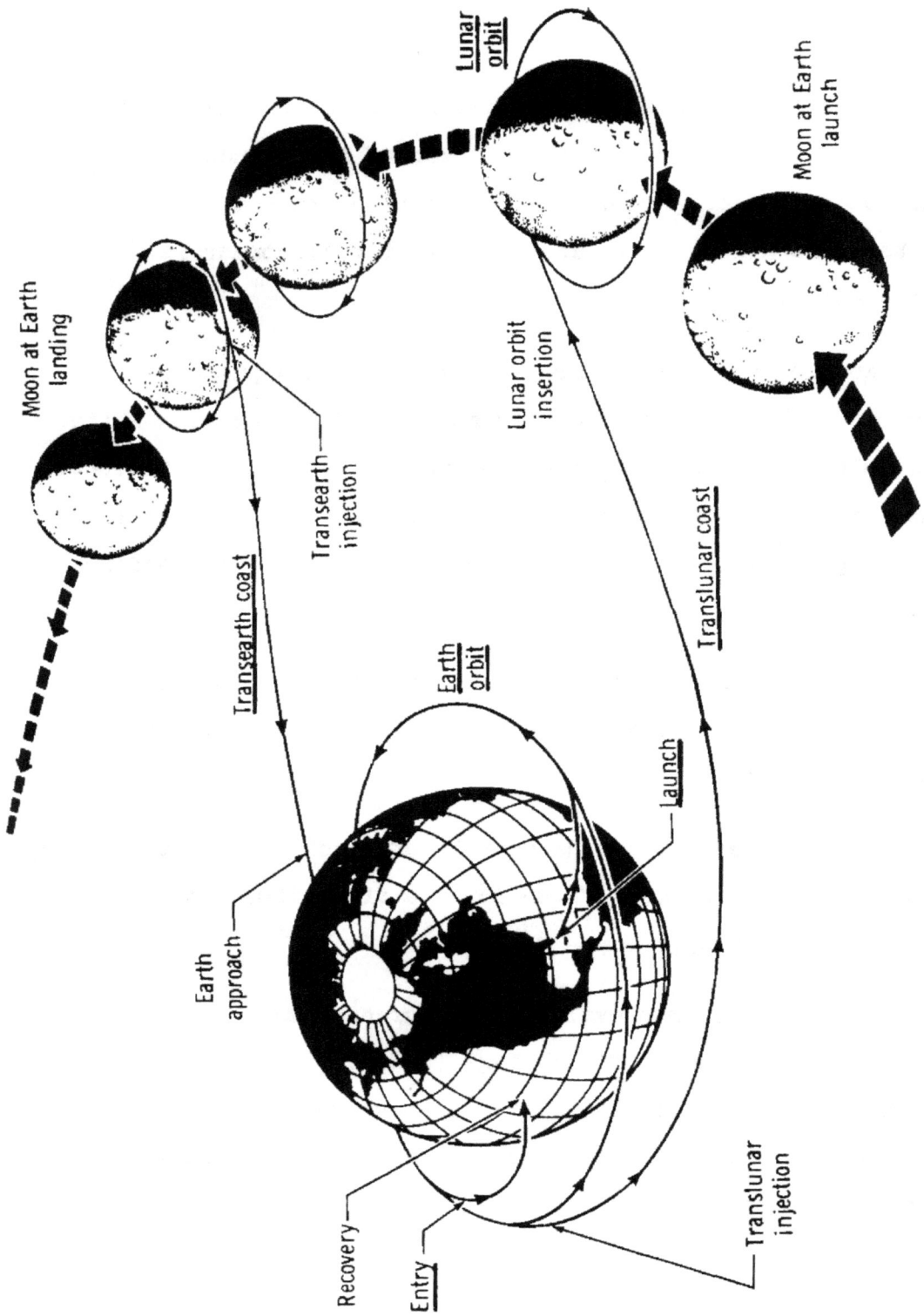

Lunar orbit

Moon at Earth landing

Moon at Earth launch

Transearth injection

Lunar orbit insertion

Transearth coast

Translunar coast

Earth approach

Earth orbit

Launch

Translunar injection

Recovery

Entry

The liquid hydrogen engine feed line for each J-2 engine has been redesigned, and the auxiliary spark igniter lines have been replaced with lines without flex joints.

A helium heater will be used as a repressurization system on the S-IVB.

The center F-1 engine on the S-IC will be cut off early to keep the acceleration forces from building up past the four "G" level.

Software changes in the instrument unit will give a new cant capability to the outboard F-1 engines. After clearing the tower, the outboard engines will cant outward two degrees to reduce the load on the spacecraft in the event of a premature cutoff of an F-1 engine.

Film cameras will not be carried on the S-II stage to record the first and second plane separations.

The forward bulkhead of the S-II fuel tank is a lightweight type that will be used on future Saturn V vehicles.

(END OF GENERAL RELEASE; BACKGROUND INFORMATION FOLLOWS)

-10a-

LES

81.9'

33'

CM

SM

52.9'

SLA

3'

IU

363'

58.5'

S-IVB

81.5'

S-II

138'

S-IC

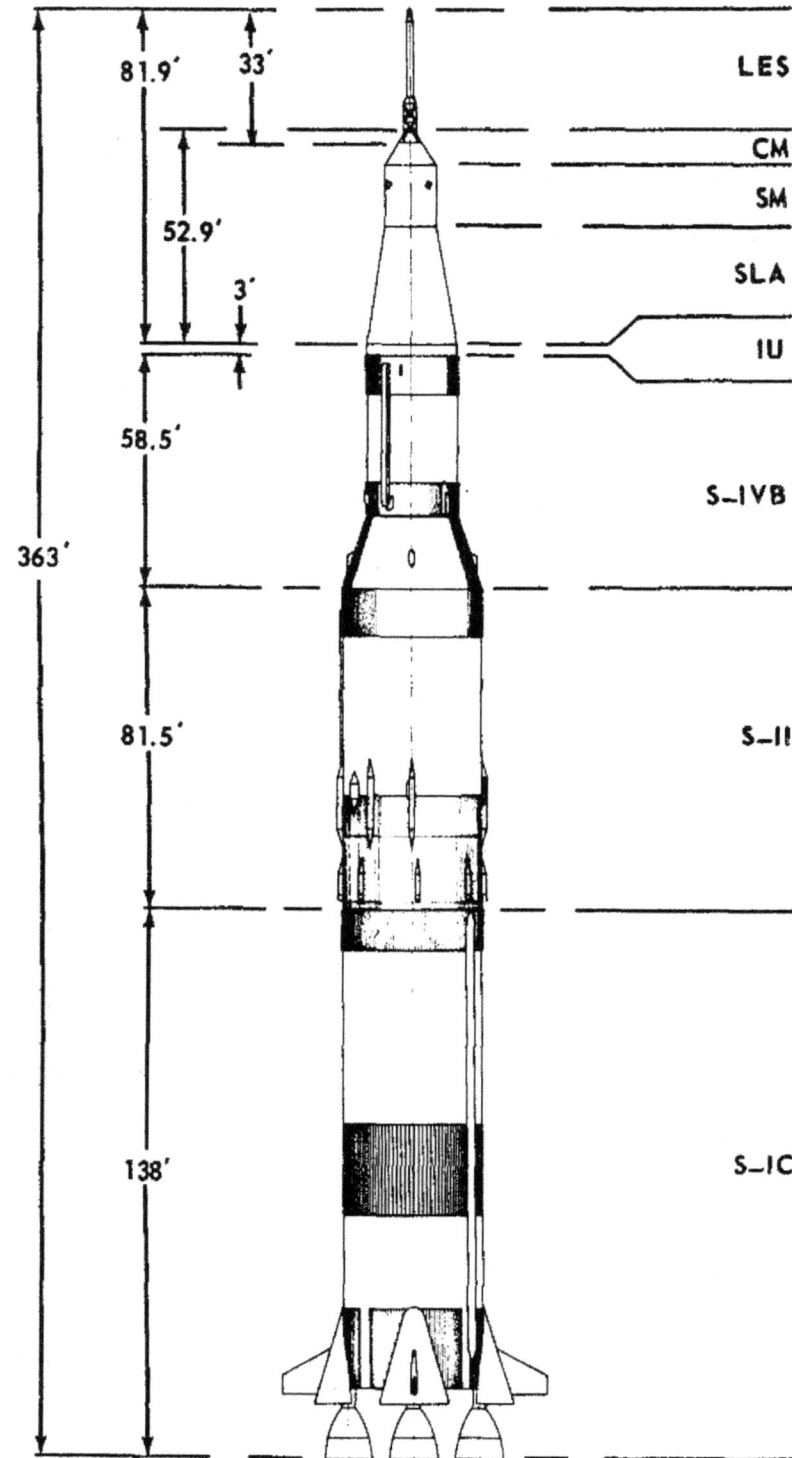

AS-503 CONFIGURATION
(APOLLO 8)

MISSION OBJECTIVES FOR APOLLO 8

* Demonstrate crew/space vehicle/mission support facilities performance during a manned Saturn V mission with command and service module.

* Demonstrate performance of nominal and selected mission activities including (1) translunar injection; (2) command service module navigation, communications and mid-course corrections; and (3) command service module consumables assesment and passive thermal control.

In addition, detailed test objectives have been designed to thoroughly wring out systems and procedures that have a direct bearing on future lunar landings and space operations in the vicinity of the Moon.

SEQUENCE OF EVENTS

NOMINAL MISSION

Time from Lift-off (Hr:Min:Sec)	Event
00:00:00	Lift-off
00:01:17	Maximum dynamic pressure
00:02:06	S-IC Center Engine Cutoff
00:02:31	S-IC Outboard Engine Cutoff
00:02:32	S-IC/S-II Separation
00:02:33	S-II Ignition
00:02:55	Camera Capsule Ejection
00:03:07	Launch Escape Tower Jettison
	Mode I/ Mode II Abort Changeover
00:08:40	S-II Cutoff
00:08:41	S-II/S-IVB Separation
00:08:44	S-IVB Ignition
00:10:06	Mode IV Capability begins
00:10:18	Mode II/Mode III Abort Changeover
00:11:32	Insertion into Earth Parking Orbit
02:50:31	Translunar Injection Ignition
02:55:43	Translunar Injection Cutoff
	Translunar Coast Begins
03:09:14	S-IVB/CSM Separation
04:44:54	Begin Maneuver to Slingshot Attitude
05:07:54	LOX Dump Begins
05:12:54	LOX Dump Ends

Time from Lift-off		Event
	TLI+ 6 Hrs.	Midcourse Correction 1
	TLI+ 25 Hrs.	Midcourse Correction 2
	LOI- 22 Hrs.	Midcourse Correction 3
	LOI- 8 Hrs.	Midcourse Correction 4
69:07:29		Lunar Orbit Insertion (LOI_1) Initiation
69:11:35		Lunar Orbit Insertion (LOI_1) Termination
73:30:53		Lunar Orbit Insertion (LOI_2) Initiation
73:31:03		Lunar Orbit Insertion (LOI_2) Termination
89:15:07		Transearth Injection Initiate
89:18:33		Transearth Injection Terminate
	TEI+ 15 Hrs.	Midcourse Correction 5
	TEI+ 30 Hrs.	Midcourse Correction 6
	EI- 2 Hrs.	Midcourse Correction 7
146:49:00		Entry Interface
147:00:00		SPLASHDOWN

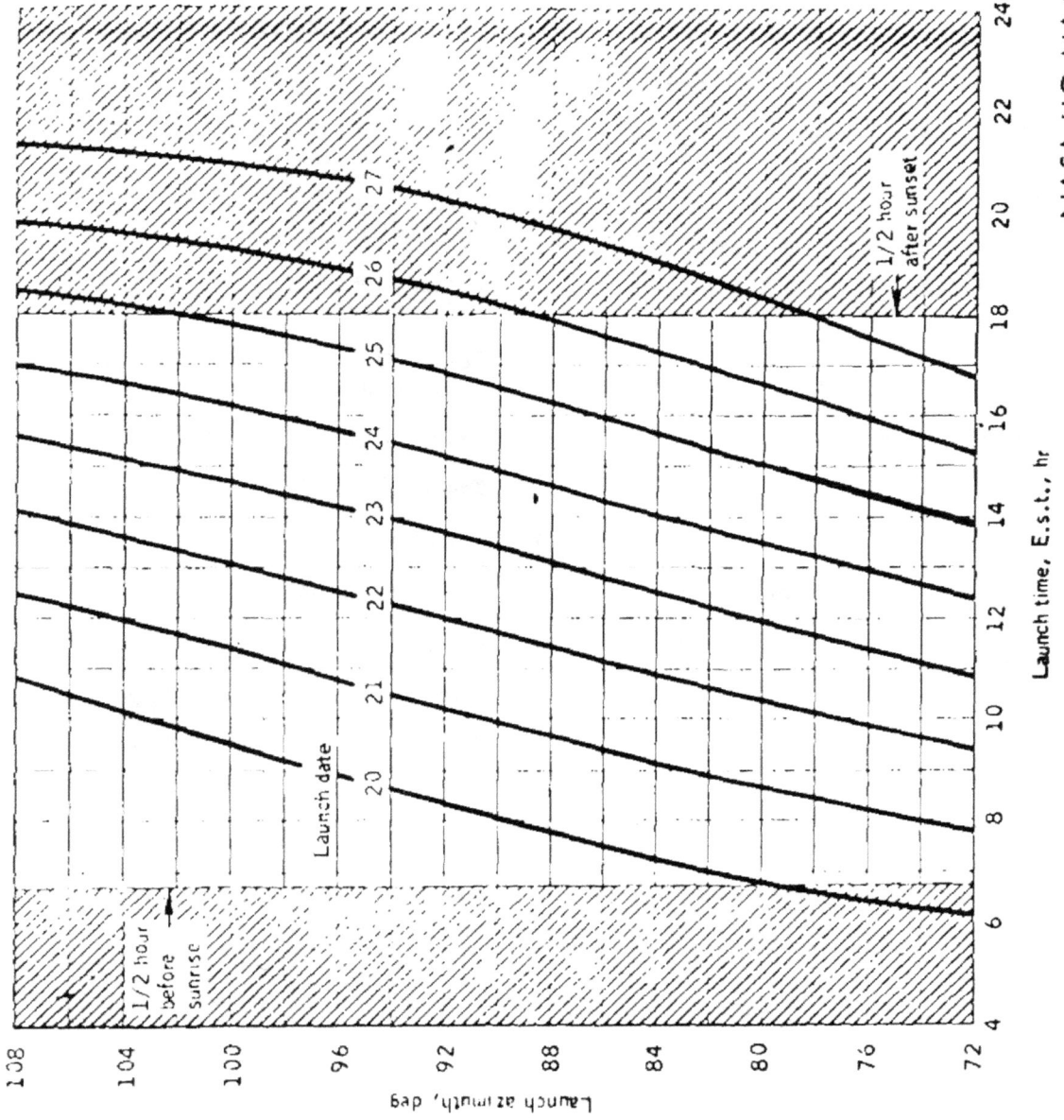

DECEMBER 1968 LAUNCH WINDOW

NASA HQ MA68-7200
10-30-68

GO/NO-GO DECISION POINTS

MISSION PHASE	TIME OF DECISION	REMARKS
LAUNCH	REAL TIME	ORBIT IS "GO" IF $H_p \geq 75$ NM
EARTH ORBIT	AFTER INSERTION	UNTIL LANDING AREA 2-1
	REV 1 U.S. PASS	TO TLI
	CRO	FOR TLI BURN
TRANSLUNAR (CONTINUOUS MONITORING)	TLI +10 -90 MINUTES	FOR ANY DISPERSIONS AND SEPARATION MANEUVER
	TLI + 2HR.	TO TLI +4HR.
	TLI + 4HR.	STILL PERFORM EARTH ORBITAL TYPE MISSION
	TLI + 5HR.	DEPENDENT UPON ΔV REQUIREMENTS FOR MID-COURSE CORRECTION: (1) CONTINUE MISSION (2) LUNAR FLYBY
	TLI + 6HR.	FOR MCC_1
	$\Delta V < 4500$ FPS TO PTP'S	FOR ANY MALFUNCTIONS THAT REQUIRE EARLY RETURNS
	TLI + 24HR.	FOR MCC_2
	LOI - 20HR.	FOR MCC_3
	LOI - 8HR.	FOR MCC_4
LUNAR ORBIT	LOI_1 - 1HR.	FOR LOI BURN: AT LEAST 4HR. LUNAR ORBIT CAPABILITY.
	LOI_2 - 1HR.	FOR LOI CIRCULARIZATION BURN. 4HR ORBIT CAPABILITY.
	TEI - 1HR.	CONTINUOUS MONITORING WHILE IN VIEW FOR TRANS EARTH INJECTION BURN.

Apollo 8 Window

Pre-Launch

Apollo 8 is scheduled to be launched from Launch Complex 39, pad A, at Cape Kennedy, Florida on December 21, 1968. The launch window opens at 7:51 a.m. EST and closes at 12:32 p.m. EST. Should holds in the launch countdown or weather require a scrub, there are six days remaining in December during which the mission could be launched.

December Launch Days for Apollo 8	WINDOW (EST)	
	Open	Close
21	7:51 a.m.	12:32 p.m.
22	9:26 a.m.	2:05 p.m.
23	10:58 a.m.	3:35 p.m.
24	12:21 p.m.	4:58 p.m.
25	1:52 p.m.	6:20 p.m.
26	3:16 p.m.	6:20 p.m.
27	4:45 p.m.	6:20 p.m.

A variable launch azimuth of 72 degrees to 108 degrees capability will be available to assure a launch on time. This is the first Apollo mission which has employed the variable launch azimuth concept. The concept is necessary to compensate for the relative positioned relationship of the Earth at launch time.

-more-

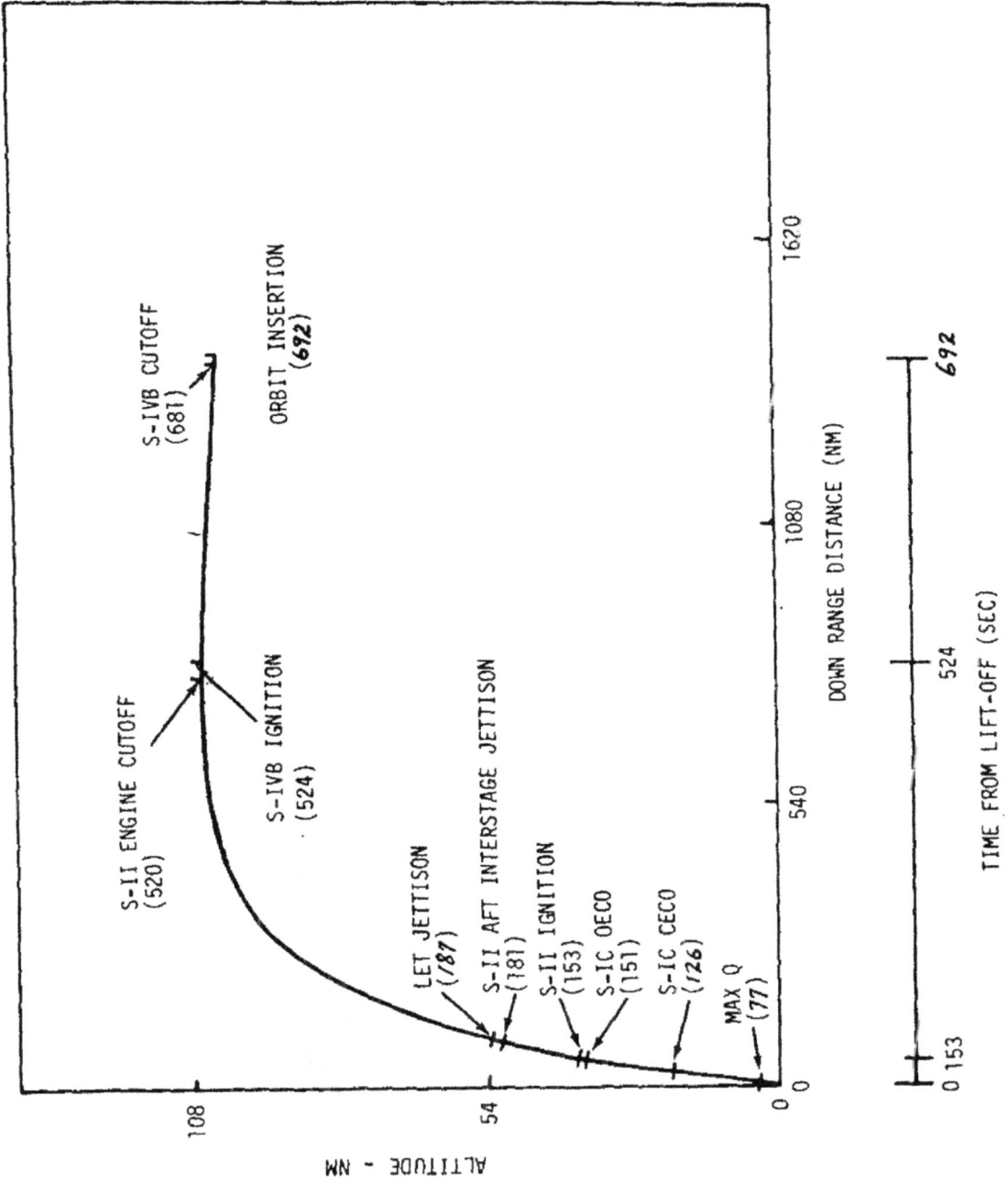

APOLLO 8 NOMINAL LAUNCH PROFILE
(72° LAUNCH AZIMUTH)

S-IVB CUTOFF
(681)

ORBIT INSERTION
(692)

S-II ENGINE CUTOFF
(520)

S-IVB IGNITION
(524)

LET JETTISON
(187)

S-II AFT INTERSTAGE JETTISON
(181)

S-II IGNITION
(153)

S-IC OECO
(151)

S-IC CECO
(126)

MAX Q
(77)

ALTITUDE - NM

108

54

0

0 153   524   1080   1620   692

DOWN RANGE DISTANCE (NM)

TIME FROM LIFT-OFF (SEC)

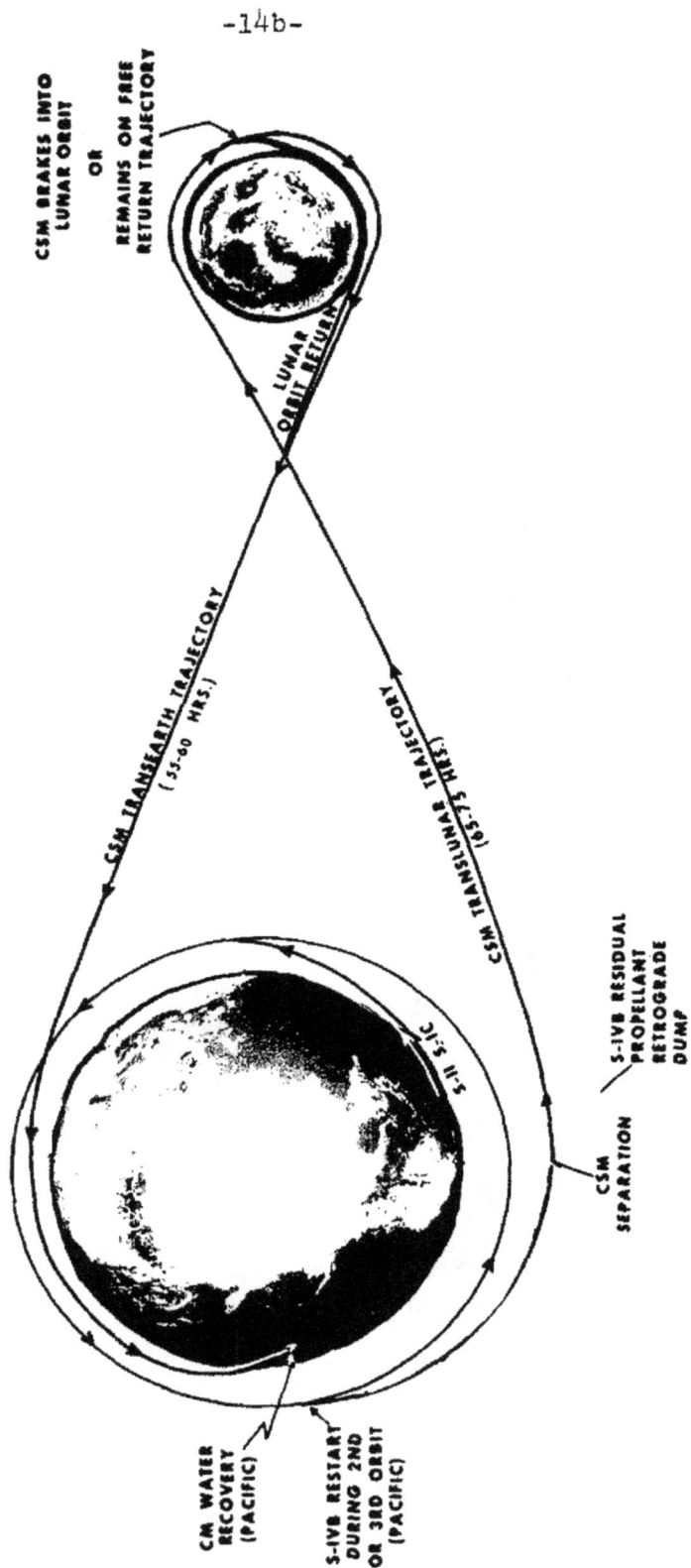

APOLLO 8 ORBIT PROFILE

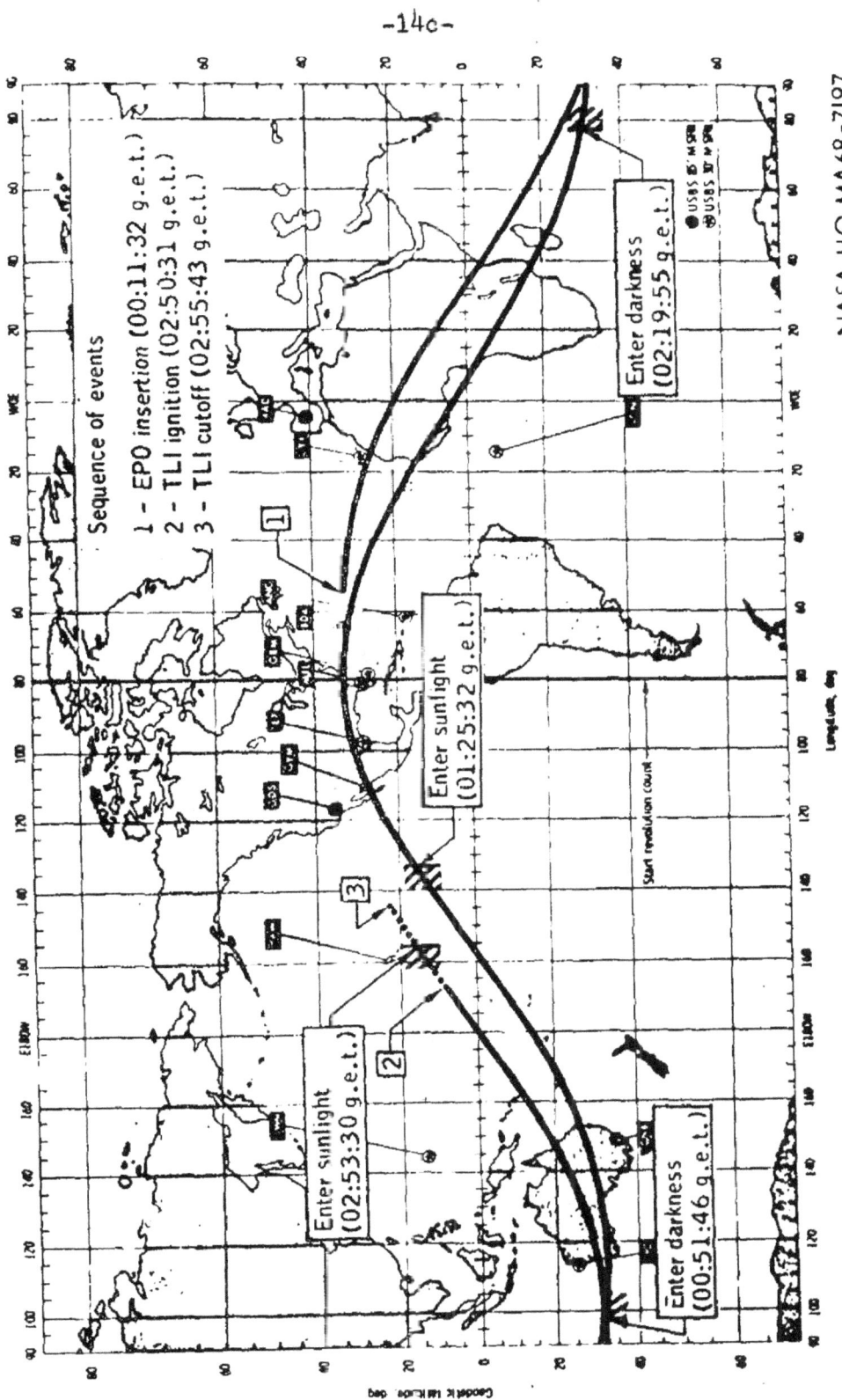

EARTH PARKING ORBIT

MISSION GROUND TRACKS

Sequence of events

1 - EPO insertion (00:11:32 g.e.t.)
2 - TLI ignition (02:50:31 g.e.t.)
3 - TLI cutoff (02:55:43 g.e.t.)

Enter darkness
(02:19:55 g.e.t.)

Enter sunlight
(01:25:32 g.e.t.)

Enter sunlight
(02:53:30 g.e.t.)

Enter darkness
(00:51:46 g.e.t.)

NASA HQ MA68-7197
10-30-68

TRANSLUNAR COAST

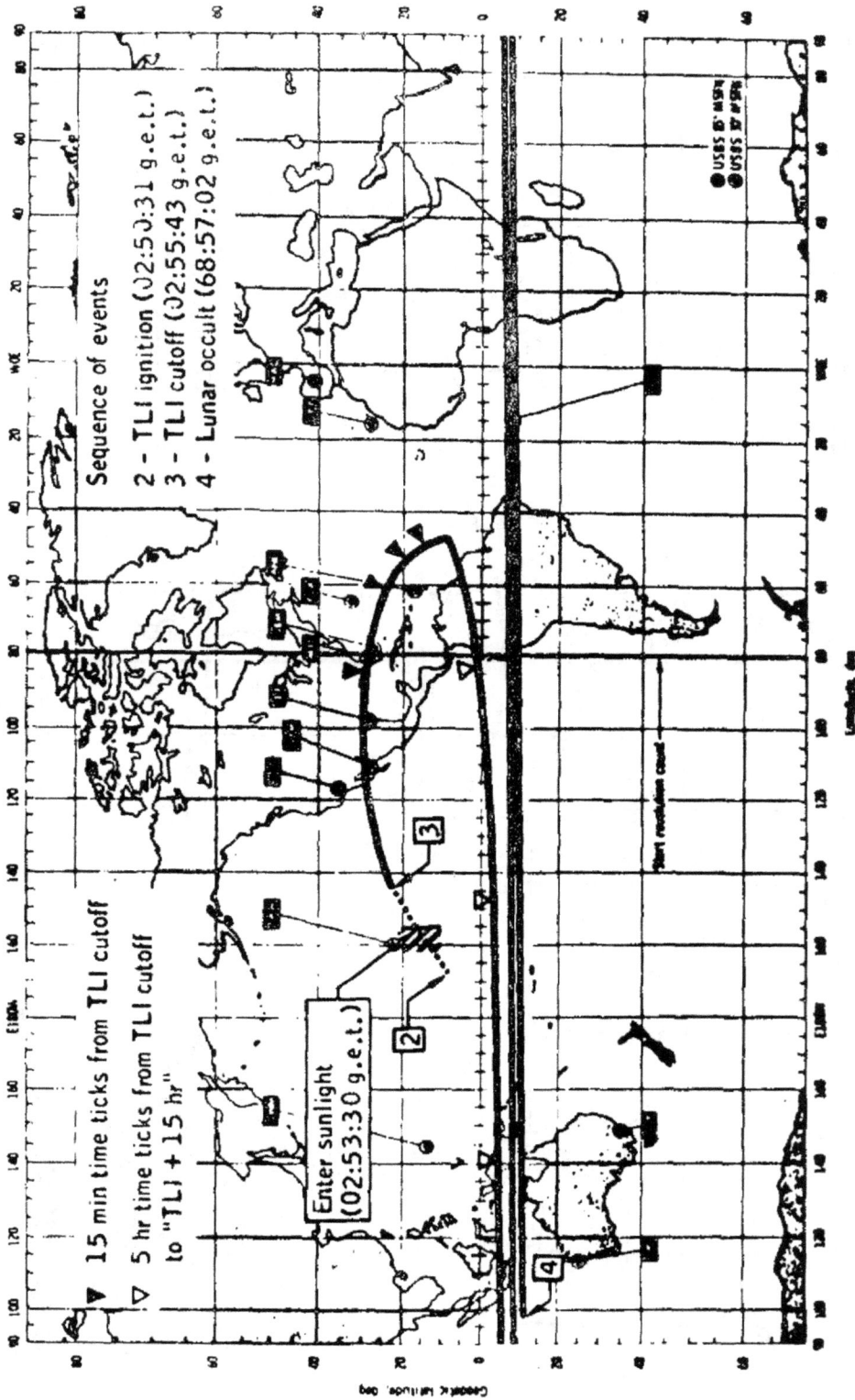

Sequence of events

2 - TLI ignition (02:50:31 g.e.t.)
3 - TLI cutoff (02:55:43 g.e.t.)
4 - Lunar occult (68:57:02 g.e.t.)

▼ 15 min time ticks from TLI cutoff

▽ 5 hr time ticks from TLI cutoff
   to "TLI + 15 hr"

Enter sunlight
(02:53:30 g.e.t.)

NASA HQ MA68-7193
10-30-68

-more-

LUNAR PARKING ORBIT

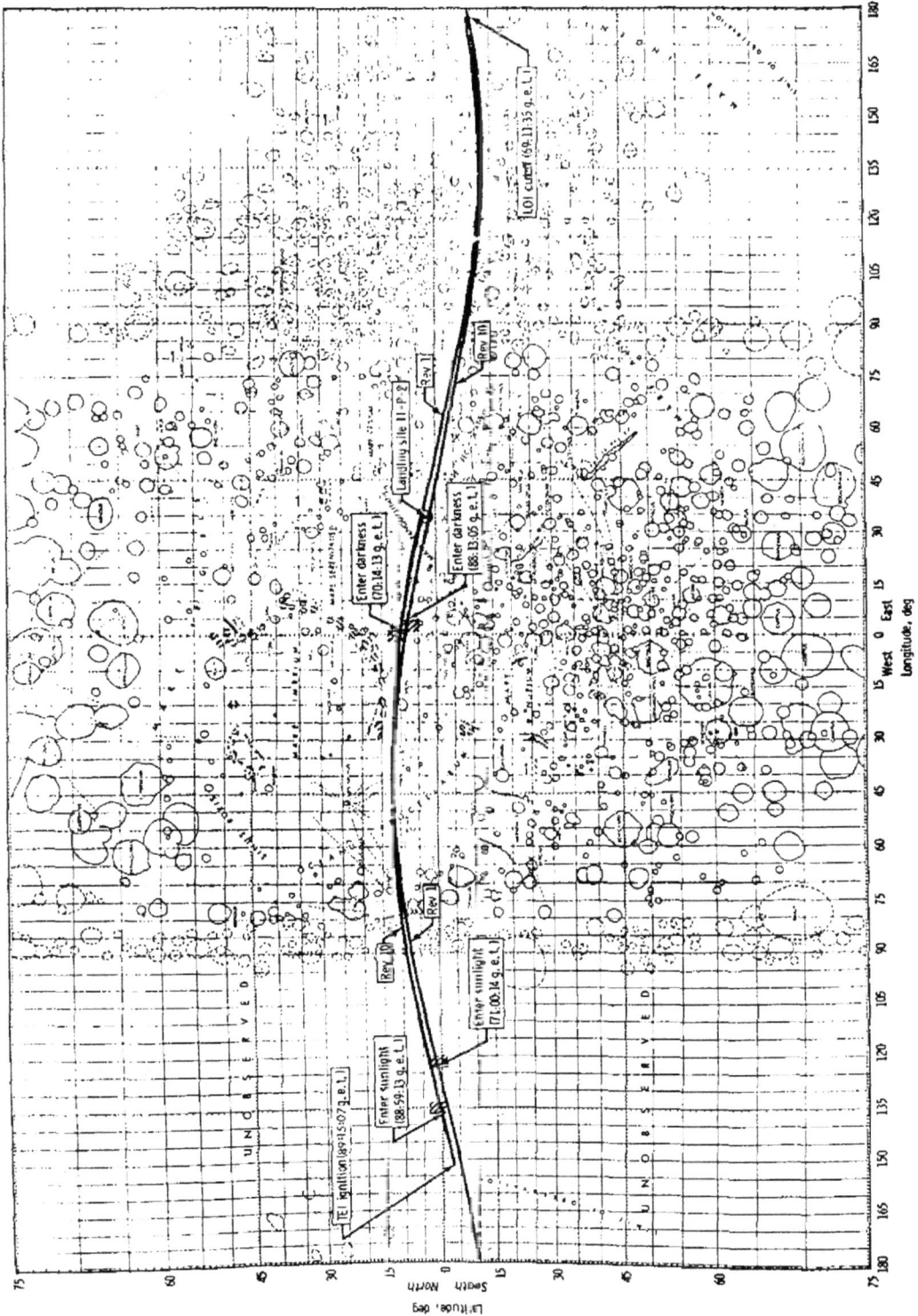

NASA HQ MA68-7192
10-30-68

TRANSEARCH COAST

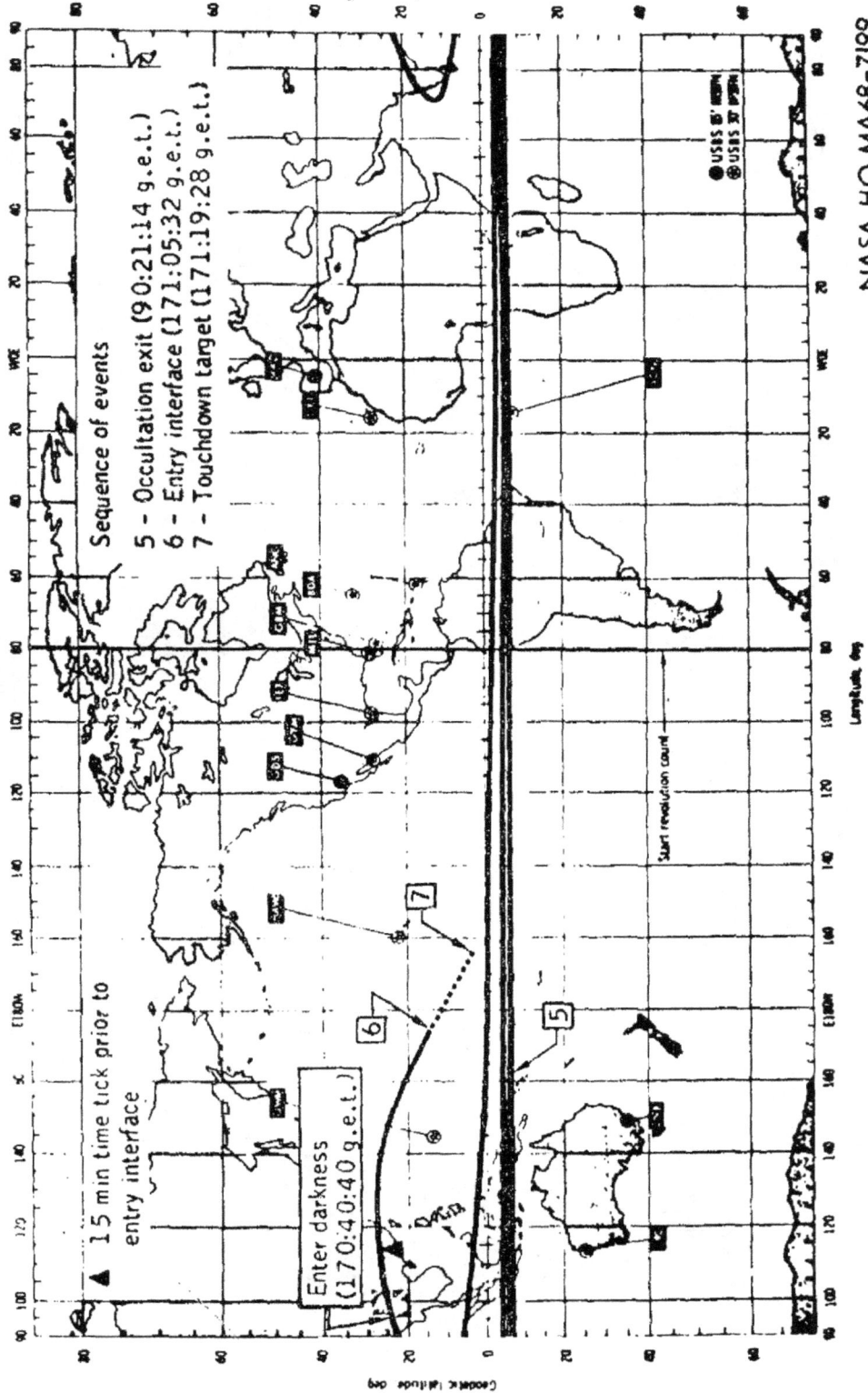

Sequence of events

5 - Occultation exit (90:21:14 g.e.t.)
6 - Entry interface (171:05:32 g.e.t.)
7 - Touchdown target (171:19:28 g.e.t.)

▲ 15 min time tick prior to entry interface

Enter darkness (170:40:40 g.e.t.)

Start revolution count

Longitude, deg

Geodetic latitude, deg

● USBS B' HORN
⊕ USBS 3T PEAN

NASA HQ MA68-7199
10-30-68

MISSION DESCRIPTION

NOTE: Information presented in this press kit is based on a nominal mission. Plans may be altered prior to or during flight to meet changing conditions.

Launch Phase

Apollo 8 will be launched from Kennedy Space Center Launch Complex 39A on a launch azimuth that can vary from 72 degrees to 108 degrees, depending upon the time of day of launch. The azimuth changes with time of day to permit a fuel optimum injection from Earth parking orbit onto a free return circumlunar trajectory. Other factors influencing the launch windows are a daylight launch (sunrise -30 min. to sunset +30 min.), and proper sun angles on lunar landmarks in the Apollo landing zone.

The planned Apollo 8 launch date of December 21 will call for liftoff time at 7:51 a.m. EST on a launch azimuth of 72 degrees. Insertion into Earth parking orbit will occur at 11 min. 32 sec. ground elapsed time (GET) at an altitude of 103 nm (119 sm, 191.3 km). The orbit resulting from this launch azimuth will have an inclination of 32.5 degrees to the equator.

Earth Parking Orbit (EPO)

Apollo 8 will remain in Earth parking orbit after insertion and will hold a local horizontal attitude during the entire period. The crew will perform spacecraft systems checks in preparation for the Translunar Injection burn.

Translunar Injection (TLI)

In the second or third revolution in Earth parking orbit, the S-IVB third stage engine will reignite over the Pacific to inject Apollo 8 toward the Moon. The velocity will increase to 35,582 feet per second (10,900 meters/sec.). Injection will begin at an altitude of 106 nm (122 sm, 197 km) while the vehicle is in darkness. Midway through the translunar injection burn, Apollo 8 will enter sunlight.

Translunar Coast

Following the translunar injection burn, Apollo 8 will spend 66 hr. 11 min. in translunar coast. The spacecraft will separate from the S-IVB stage about 20 minutes after the start of the translunar injection burn, using the service module Reaction Control System (RCS) thrusters to maneuver out to about 50 to 70 feet from the stage for a 13-minute period of station keeping. The spacecraft will move away some five minutes later in an "evasive maneuver" while the S-IVB is commanded to dump residual liquid oxygen (LOX) through the J-2 engine bell about 1 hr. and 30 min. after separation. The third stage auxiliary propulsion system will be operated to depletion. The LOX dumping is expected to impart a velocity of about 90 fps to the S-IVB to: lessen probability of recontact with the spacecraft; and place the stage in a "slingshot" trajectory passing behind the Moon's trailing edge and on into solar orbit.

Four midcourse correction burns are possible during the translunar coast phase, depending upon the accuracy of the trajectory. The first burn, at translunar injection +6 hrs., will be done if the needed velocity change is greater than 3 fps; the second at TLI +25 hrs.; the third at lunar orbit insertion (LOI) -22 hrs., and the fourth at LOI -8 hrs. The last three burns will be made only if the needed velocity is greater than 1 fps.

Lunar Orbit Insertion (LOI)

The first of two lunar orbit insertion burns will be made at 69:07:29 GET at an altitude above the Moon 69 nm (79 sm, 126.8 km). LOI No. 1 will have a nominal retrograde velocity change of 2,991 fps (912 m/sec) and will insert Apollo 8 into a 60 x 170 nm (69 x 196 sm, 111 x 314.8 km) elliptical lunar parking orbit. At 73:30:53, LOI Burn No. 2 will circularize the lunar parking orbit at 60 nm (69 sm, 111 km) with a retrograde velocity change of 138 fps (42.2 m/sec). The lunar parking orbit will have an inclination of 12 degrees to the lunar equator.

Lunar Parking Orbit (LPO)

During the 10 revolutions of lunar parking orbit, the Apollo 8 crew will perform lunar landmark tracking and Apollo landing site tracking and photographic tasks, and stereo photography of the lunar surface from terminator to terminator. The last two lunar revolutions will be spent in preparation for the transearth injection burn.

LUNAR ORBIT ACTIVITIES

Revolution	Activity
1	LOI_1
2	Housekeeping, Systems Checks, and Landmarks sightings
3	LOI_2, System Checks and Training Photography
4,5,6,7	General landmark sightings, Stereo photography, and general landmark photography
8	General photography, Landmark sightings, Solar corona, Dim Sky, and Earth Shine photography
9	Prepare for TEI and perform oblique stereo strip photography
10	Perform TEI

Trans-Earth Injection (TEI)

The SPS trans-Earth injection burn is nominally planned
for 89:15:07 GET with a posigrade velocity increase of 3520
fps (1073 m/sec). The burn begins on the backside of the
Moon and injects the spacecraft on a trajectory toward the
Earth. It will reach 400,000 feet altitude above Earth at
146:49:00 GET.

APOLLO 8
LUNAR SEQUENCE OF EVENTS

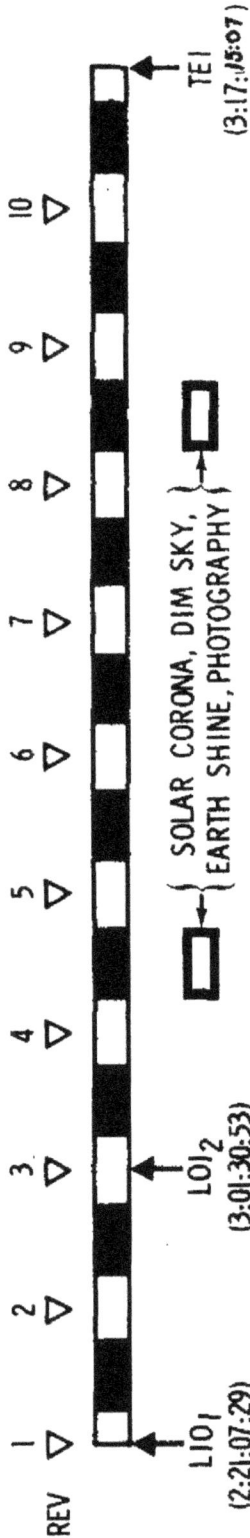

REV   1 ▷   2 ▷   3 ▷   4 ▷   5 ▷   6 ▷   7 ▷   8 ▷   9 ▷   10 ▷

LIO₁
(2:21:07:29)

LOI₂
(3:01:30:53)

TEI
(3:17:15:07)

{ SOLAR CORONA, DIM SKY,
EARTH SHINE, PHOTOGRAPHY }

SEQUENCE PHOTO'S ▯

VERTICAL STEREO ▯

LIGHTING EVALUATION ▮

LANDMARK (LM) TRACKING
FOR DESCENT TARGETING   I

UNKNOWN LM (ULM) TRACK :
(INCLUDES PHOTOGRAPHY)   II

IV   20⁰ AFTER EXIT FROM DARK   30⁰ BEFORE SUB-SOLAR

III   30⁰ AFTER S-SOLAR

OBLIQUE STEREO ▯

* TIME IN GET BEGINNING AT END OF DAY 2.   PENUMBRA
IS 13-15 SECONDS DURATION PRIOR TO AND AFTER UMBRA.

UMBRA* SCHEDULE		
REV	ENTER	EXIT
1	2:22:14:13	2:23:00:14
2	3:00:22:47	3:01:08:46
3	02:22:57	03:08:53
4	04:21:29	05:07:25
5	06:20:02	07:05:59
6	08:18:44	09:04:39
7	10:17:17	11:03:13
8	12:15:51	13:01:47
9	14:14:31	15:00:26
10	16:13:05	16:59:13

Transearth Coast, Midcourse Manuevers

During the approximate 57-hour Earth return trajectory, the Apollo 8 crew will perform navigation sightings on stars, and lunar and Earth landmarks, communications tests and spacecraft passive thermal control tests. Three midcourse corrections are possible during the transearth coast phase, and their values will be computed in real time. The midcourse corrections, if needed, will be made at transearth injection +15 hr., TEI +30 hr., and entry interface -2 hr. (400,000 ft. altitude).

Entry, Landing

Apollo 8 command module will be pyrotechnically separated from the service module approximately 15 minutes prior to reaching 400,000 ft. altitude. Entry will begin at 146:49:00 GET at a spacecraft velocity of 36,219 fps (11005 m/sec). The crew will fly the entry phase with the G&N system to produce a constant deceleration (average 4 Gs) for a direct entry, rather than the dual-pulse "skip" entry technique considered earlier in Apollo program planning. Splashdown is targeted for the Pacific Ocean at 165 degrees West longitude by 4 degrees 55 min. North latitude. The landing footprint will extend some 1350 nm (1560 sm, 2497 km) from its entry point. Splashdown will be at 13 min. 46 sec. after entry.

CM END-OF-MISSION ENTRY AND LANDING POINTS*

Day Of Launch	Entry Point Latitude	Entry Point Longitude	Landing Point Latitude	Landing Point Longitude
21 Dec	14°42'N	174°30'E	4°55'N	165°00'W
22 Dec	5°35'N	173°50'E	1°00'S	165°00'W
23 Dec	1°20'N	174°35'E	8°10'S	165°00'W
24 Dec	10°15'S	172°15'E	12°50'S	165°00'W
25 Dec	18°55'S	171°25'E	18°00'S	165°00'W
26 Dec	25°00'S	170°45'E	22°10'S	165°00'W
27 Dec	22°25'S	170°55'E	25°25'S	165°00'W

*These points are for a 72 degree launch azimuth. Other launch azimuths will change the data slightly.

GEODETIC ALTITUDE VERSUS RANGE TO GO

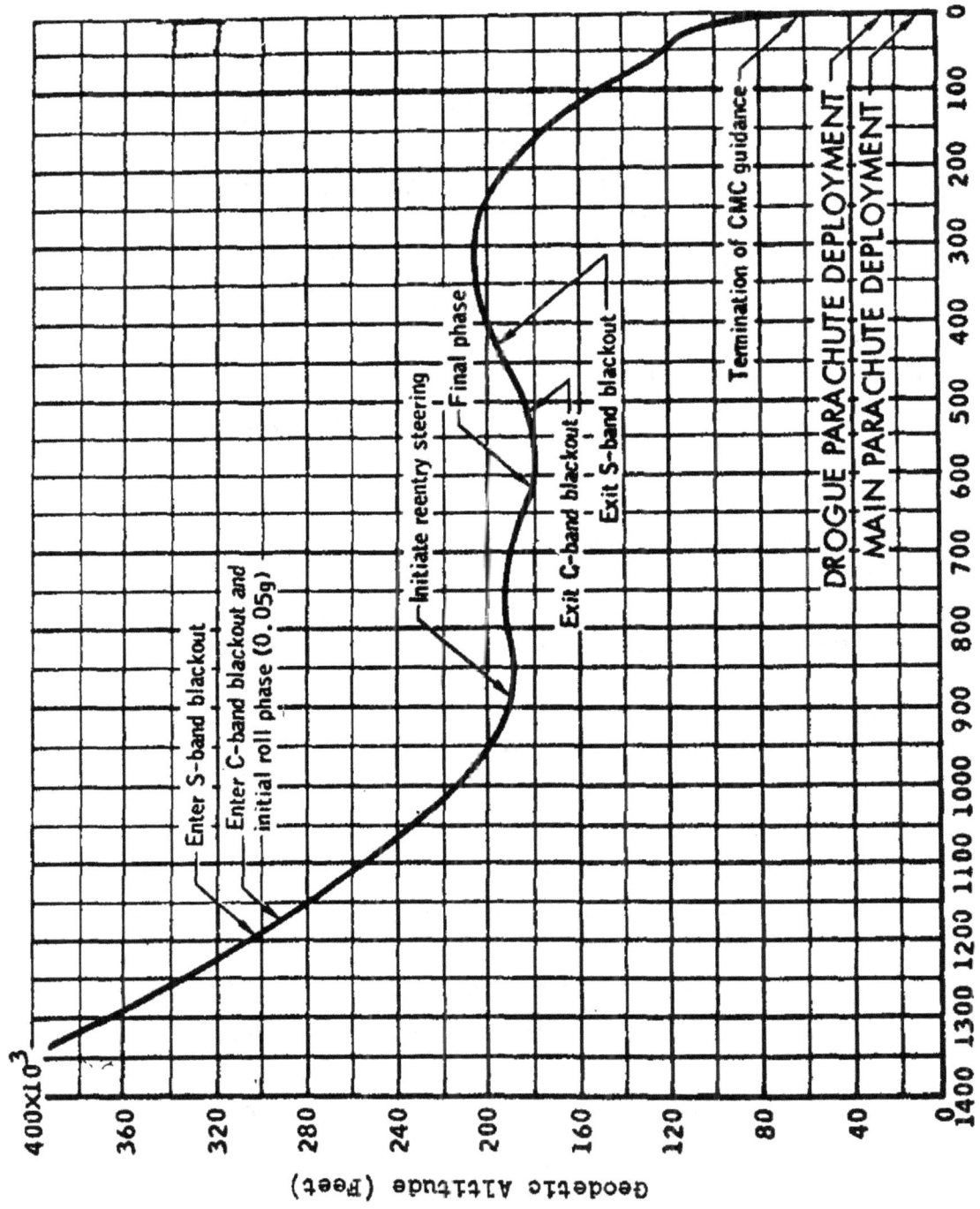

Enter S-band blackout

Enter C-band blackout and initial roll phase (0.05g)

Initiate reentry steering

Final phase

Exit C-band blackout

Exit S-band blackout

Termination of CMC guidance

DROGUE PARACHUTE DEPLOYMENT

MAIN PARACHUTE DEPLOYMENT

Geodetic Altitude (Feet)

400x10³ 360 320 280 240 200 160 120 80 40 0

1400 1300 1200 1100 1000 900 800 700 600 500 400 300 200 100 0

Range to go, n. mi.

NASA HQ MA68-7196
10-30-68

FLIGHT PLAN

Crew Activities

The Apollo 8 flight plan calls for at least one crewman to be awake at all times. The normal work/rest cycle will be 17 hours of work followed by seven hours of rest. The command module pilot and lunar module pilot sleep periods are scheduled simultaneously.

Following the translunar injection burn, the crew will take off their space suits and put on the inflight coveralls for the duration of the mission.

The Apollo 8 spacecraft normally will remain fully powered up throughout the entire mission, with the stabilization and control system and navigation sextant and scanning telescope turned on as needed. The inertial measuring unit and command module computer will stay in the "operate" mode.

Changes of lithium hydroxide canisters for absorbing cabin carbon dioxide are scheduled at times when all crewmen are awake, and all scheduled maneuvers will be made when all crewmen are awake.

Flight plan updates will be relayed to the Apollo 8 crew over S-Band frequencies each day for the coming day's activities.

Following is a brief summary of tasks to be accomplished in Apollo 8 on a day-to-day schedule. The tasks are subject to changes to suit opportunity or operational factors:

Launch Day (0-24 hours):

* CSM systems checkout following Earth orbit insertion

* Translunar injection burn, separation from S-IVB and transposition maneuver

* Monitor S-IVB LOX blowdown and "slingshot" maneuver

* Five sets of translunar coast star-Earth horizon navigation sightings

* Perform first midcourse correction burn if required

* Star-Earth landmark navigation sightings

Second Day (24-48 hours):

* Midcourse correction burns Nos. 2 and 3 if required

* Star-Earth horizon, star-lunar horizon navigation sightings

    Third Day (48-72 hours):

* Star-lunar horizon navigation sightings

* Midcourse correction burn No. 4 if required

* Lunar orbit insertion burn into initial 60 x 170 nm
  (69 x 196 sm, 111 x 314.8 km) orbit

* General lunar landmark observation, photography, onboard
  television

* Preparations for lunar orbit circularization burn

* Align IMU once during each lunar orbit dark period

    Fourth Day (72-96 hours):

* Circularize lunar orbit to 60 nm (69 sm, 111 km)

* Vertical stereo, convergent stereo navigation photography

* Solar corona photography

* Landmark and landing site tracking and photography

* Star-lunar landmark navigation sightings

* Transearth injection burn

    Fifth Day (96-120 hours):

* Star-lunar horizon, star-Earth horizon navigation sightings

* Midcourse correction burns Nos. 5 and 6 if required

    Sixth Day (120-144 hours):

* Star-Earth horizon and star-lunar horizon navigation
  sightings

* Midcourse correction burn No. 7 if required

    Seventh Day (144 hours to splashdown):

* Star-Earth horizon and star-Earth landmark navigation
  sightings

* CM/SM separation, entry and splashdown.

Recovery Operations

The primary recovery line for Apollo 8 is in the mid-Pacific along the 165th west meridian of longitude where the primary recovery vessel, the aircraft carrier USS Yorktown will be on station. Nominal splashdown for a full-duration lunar orbit mission launched on time December 21 will be at 4 degrees 55 minutes north x 165 degrees west at a ground elapsed time of 147 hours.

Other planned recovery lines for a deep-space mission are the East Pacific line extending parallel to the coast-line of North and South America, the Atlantic Ocean line running along the 30th West meridian in the northern hemi-sphere and along the 25th West meridian in the southern hemisphere, the Indian Ocean line extending along the 65th East meridian, and the West Pacific line along the 150th East meridian in the northern hemisphere and jogging to the 170th East meridian in the southern hemisphere. Secondary landing areas for a possible Earth orbital alternate mission have been established in two zones in the Pacific and two in the Atlantic.

Ships on station in the launch abort area stretching 3,400 miles eastward from Cape Kennedy include the Helicopter landing platform USS Guadalcanal, one of whose duties will be retrieval of camera cassettes from the S-IC stage; the transport USS Rankin, the tracking ship USNS Vanguard which will be released from recovery duty after insertion into Earth parking orbit, and the oiler USS Chuckawan.

In addition to surface vessels deployed in the launch abort area and the primary recovery vessel in the Pacific, 16 HC-130 aircraft will be on standby at eight staging bases around the Earth: Tachikawa, Japan; Pago Pago, Samoa; Hawaii; Bermuda; Lajes, Azores; Ascension Island; Mauritius, and Panama Canal Zone.

Apollo 8 recovery operations will be directed from the Recovery Operations Control Room in the Mission Control Center and will be supported by the Atlantic Recovery Control Center, Norfolk, Va.; Pacific Recovery Control Center, Kunia, Hawaii; and control centers at Ramstein, Germany; and Albrook AFB, Canal Zone.

The Apollo 8 crew will be flown from the primary
recovery vessel to Manned Spacecraft Center after recovery.
The spacecraft will receive a preliminary examination, safing
and power-down aboard the Yorktown prior to offloading at Ford
Island, Hawaii, where the spacecraft will undergo a more complete
deactivation. It is anticipated that the spacecraft will be
flown from Ford Island to Long Beach, Calif., within 72 hours,
and thence trucked to the North American Rockwell plant in
Downey, Calif., for postflight analysis.

-more-

APOLLO 8
RECOVERY AREAS

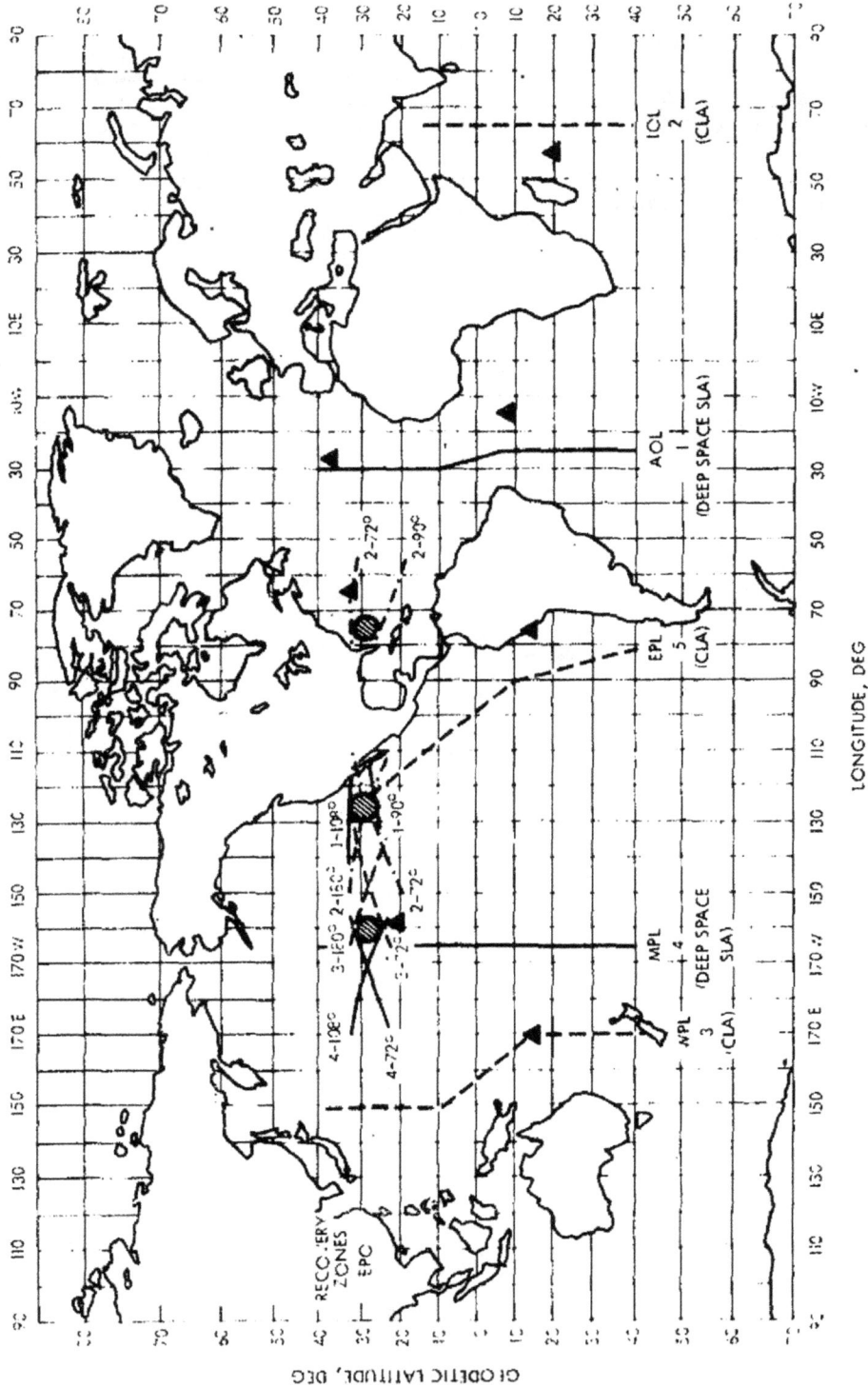

NASA HQ MA68-713
10-15-68

REENTRY COVERAGE

DECEMBER 21,1968-LAUNCH

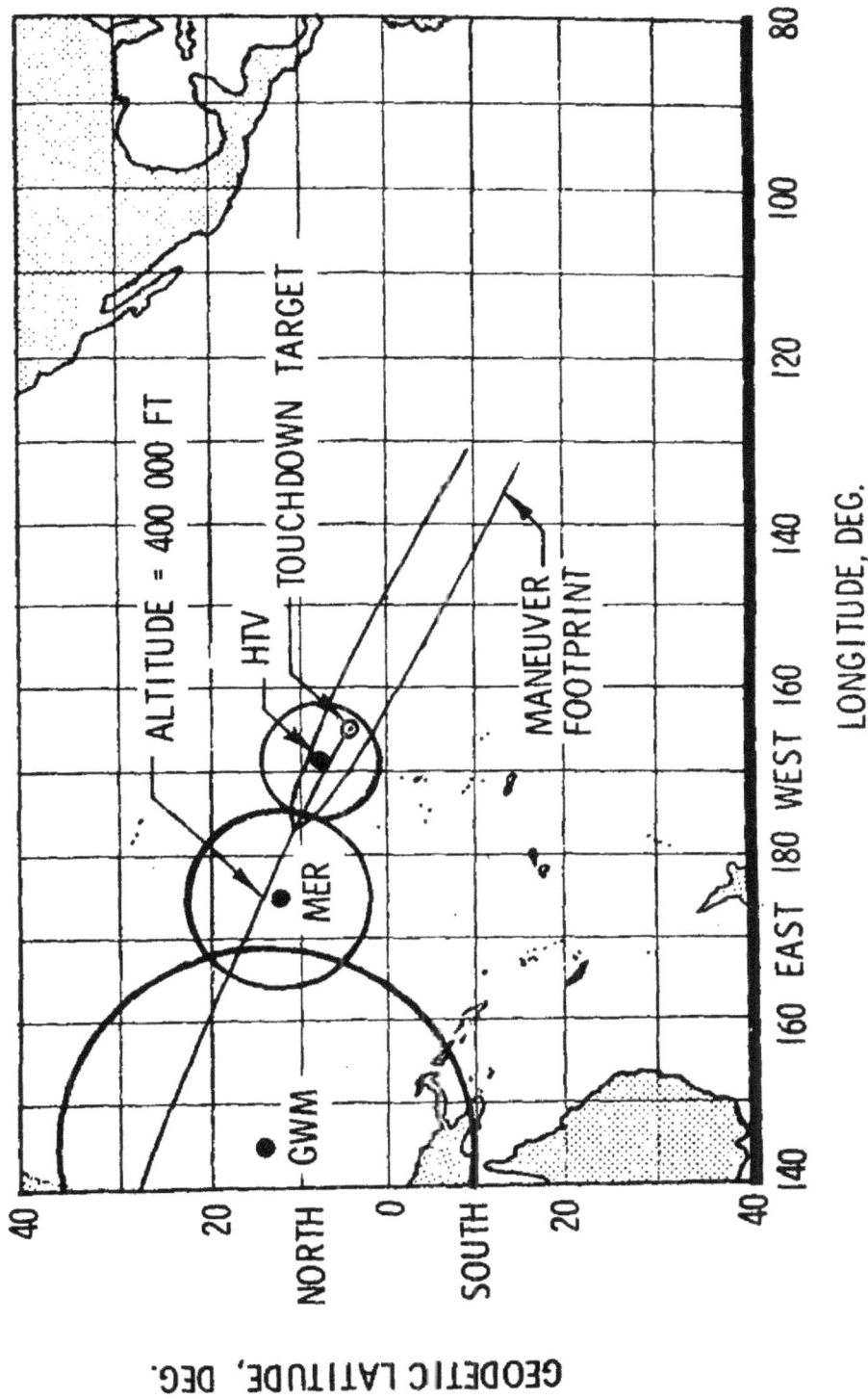

ALTITUDE = 400 000 FT

HTV — TOUCHDOWN TARGET

MANEUVER FOOTPRINT

MER

GWM

GEODETIC LATITUDE, DEG.

NORTH 40 20 0

SOUTH 20 40

140 160 EAST 180 WEST 160 140 120 100 80

LONGITUDE, DEG.

NASA HQ MO68-7242
11-8-68

APOLLO 8 PRIMARY LANDING AREA
AND FORCE DEPLOYMENT

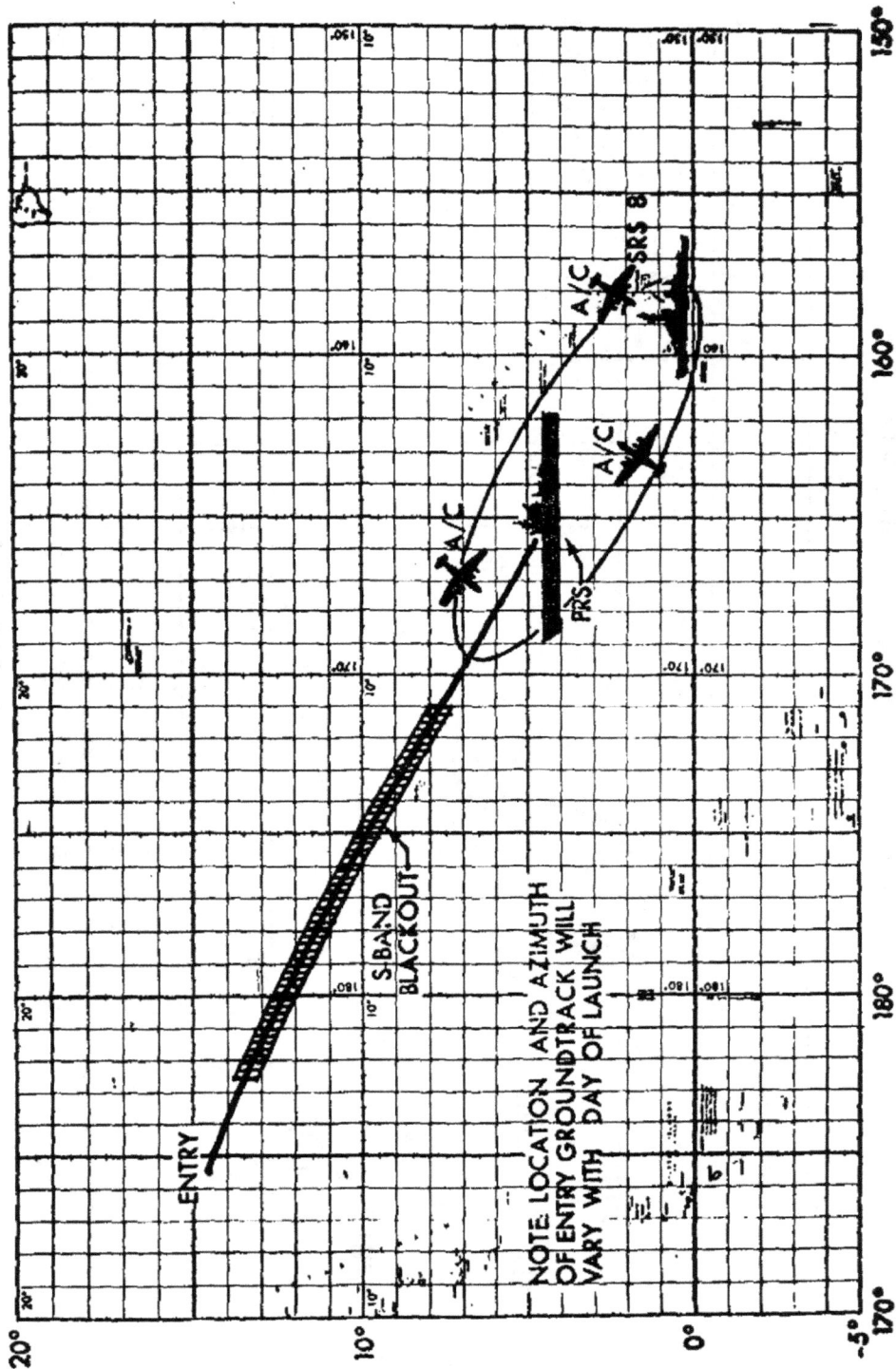

NASA HQ MA68-7312
11-8-68

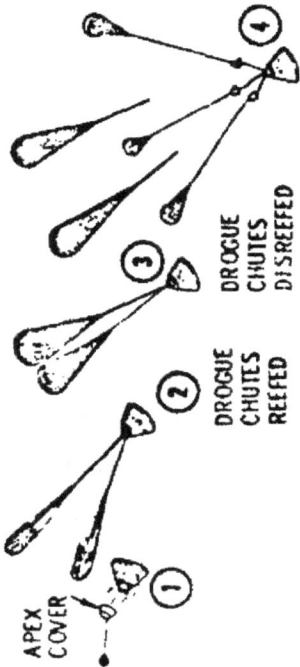
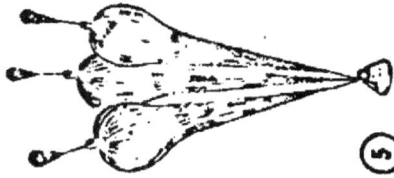
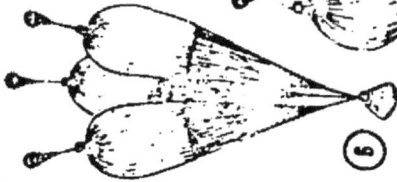
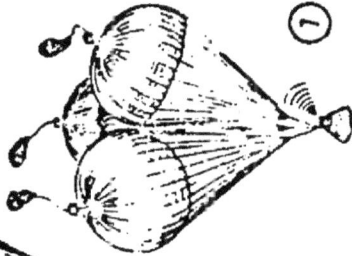

APEX COVER

DROGUE CHUTES REEFED

DROGUE CHUTES DISREEFED

MAIN CHUTES DISREEFED

1. APEX COVER JETTISONED AT 24,000 FT + 4 SEC (TLM)
2. DROGUE CHUTES DEPLOYED REEFED AT 24,000 FT +2 SEC (TLM)
3. DROGUE CHUTE SINGLE STAGE DISREEF 10 SEC
4. MAIN CHUTE DEPLOYED REEFED VIA PILOT CHUTES AND DROGUE CHUTES RELEASED AT 10,000 FT (TLM)
5. MAIN CHUTE INITIAL INFLATION
6. MAIN CHUTE FIRST STAGE DISREEF 6 SEC
7. VHF RECOVERY ANTENNAS AND FLASHING BEACON DEPLOYED 8 SEC
8. MAIN CHUTE SECOND STAGE DISREEF 10 SEC
9. MAIN CHUTES RELEASED & LM PRESS PYRO VALVE CLOSED AFTER SPLASH DOWN (TLM)

SPLASH DOWN VELOCITIES:
3 CHUTES - 31 FT/SEC
2 CHUTES - 36 FT/SEC

Earth Landing System, Normal Sequence

ALTERNATE MISSIONS

Several alternate mission plans have been prepared for the Apollo 8 mission, and the scope of the alternate is dependent upon when in the mission timeline it becomes necessary to switch to the alternate.

For example, if there is an early shutdown of the S-IVB stage during the Earth parking orbit insertion burn, the service propulsion system engine would be ignited for a contingency orbit insertion (COI), and again ingited later in the mission to boost the spacecraft to a 4,000 nm (4,610 sm, 7,400 km) apogee. The second burn into the high ellipse would be done only if the COI burn required less than 900 fps (274.5 m/sec) velocity increase. The COI-high apogee mission is Alternate 1, and would have a duration of up to 10 days.

Alternate mission 2 would be followed if the S-IVB failed to restart for the translunar injection burn out of Earth parking orbit. This alternate calls for the SPS engine to boost the spacecraft to the 4,000-nm (4,610 sm, 7,400 km) apogee for two to four revolutions. A deboost maneuver later would lower apogee and the mission would continue in low Earth orbit for 10 days.

Alternate mission 3 is split into three subalternates, each depending upon the apogee that can be reached after early S-IVB cutoff during the translunar injection burn. In the 3a alternate where apogee would be between 100 to 1,000 nm, (115-1,152 sm, 185-1,853 km) the orbit would be tuned up with an SPS burn to permit landmark sighting and the mission would follow the alternate 2 timeline. If apogee ranged between 1,000 nm (1,152 sm, 1853 km) and 25,000 nm (28,800 sm, 46,250 km), a phasing maneuver would be made at first perigee to shift a later perigee over a network station, where a deboost burn would lower apogee to 400 nm (461 sm, 740 km) and the mission would continue in low Earth orbit for the 3b alternate.

Alternate 3c would be followed if apogee was between 25,000 nm (28,800 sm, 46,250) and 60,000 nm (69,100 sm, 111,000 km). This alternate calls for a phasing maneuver at first perigee to shift later perigee to the recovery area. A second maneuver at the later perigee would adjust the elliptical orbit to one with a semi-synchronous period of about 12 hours--that is, there would be two daily perigee deorbit periods, one over the Pacific and one over the Atlantic. The entire mission would be flown in this type orbit, including direct entry from the high ellipse.

Early TLI cutoff which would produce an apogee greater than 60,000 nm (69,100 sm, 111,000 km) would fall into the alternate 3d category. This alternate calls for a circumlunar flyby using the Service Propulsion System to correct the flight profile back to a free-return trajectory. In some cases a lunar orbit mission may be possible.

All alternate mission plans call for water landings along the nominal Pacific recovery line or in the Atlantic and in general follow the lunar orbit mission timeline. Entry velocities from any of the alternates range between 26,000 fps and 36,000 fps (7,930-11,080 m/sec).

Apollo 8 alternate mission are summarized in the table below:

Apollo 8 Alternate Missions

Condition	Summary Alternate Plan
1. S-IVB early cutoff on EPO burn, COI with SPS	If COI takes less than 900 fps (274.5 m/sec), use SPS to raise apogee to 4,000 nm (4,610 sm, 7,200 km); if COI takes more than 900 fps (274.5 m/sec) remain in Earth orbit up to 10 days.
2. In EPO but S-IVB fails to restart for TLI	Use SPS to raise apogee to 4,000 nm (4,610 sm, 7,400 km); remain in high ellipse for 2-4 revolutions, then deboost to low Earth orbit for remainder of 10 days.
3. Early S-IVB TLI cutoff producing apogee of:	
a. 100-1,000 nm (115-1,152 sm, 185-1,853 km)	Burn to 4,000 nm (4,610 sm, 7,400 km) apogee, phase adjust for landmark sightings, remain in high ellipse for 2-4 revolutions, lower apogee and continue low Earth orbit mission.
b. 1,000-25,000 nm (1,152-28,800 sm, 1,853-46,250 km)	Make phasing maneuver at first perigee to shift later perigee over network station; at that perigee lower apogee to 400 nm (461 sm, 740 km); later SPS burn lowers apogee further and mission continues in low Earth orbit.

c. 25,000-60,000 nm (28,800-69,100 sm, 46,250-111,000 km)

Remain in established trajectory and make direct entry. (SPS fuel remaining not enough to lower apogee to 400 nm (461 sm, 740 km) and still perform deorbit burn)

d. More than 60,000 nm (69,100 sm, 111,000 km)

Correct trajectory to lunar flyby and Earth free-return with SPS, direct entry.

ABORT MODES

The Apollo 8 mission can be aborted at anytime during the launch phase or during later phases after a successful insertion into earth orbit.

Abort modes can be summarized as follows:

Launch phase--

Mode I -- Launch escape tower propels command module safely away from launch vehicle. This mode is in effect from about T-30 min. when LES is armed until LES jettison at 3:07 GET and command module landing point can range from the Launch Complex 39A area to 520 nm (600 sm, 964 km) downrange.

Mode II - Begins when LES is jettisoned and runs to 10:00 GET. Command module separates from launch vehicle and free-falls in a full-lift entry with landing between 400 and 3200 nm (461-3680 sm, 741-5930 km) downrange.

Mode III - Begins when full-lift landing point reaches 3200 nm (3680 sm, 5930 km) and extends through orbital insertion. The CSM would separate from the launch vehicle, and if necessary, an SPS retrograde burn would be made, and the command module would be flown half-lift to entry and landing between 3000 and 3350 nm (3450-3850 sm, 5560-6200 km) downrange.

Mode IV and Apogee Kick - Begins after the point the SPS could be used to insert the CSM into an earth parking orbit ---from about 10 minutes after liftoff. The SPS burn into orbit would be made two minutes after separation from the S-IVB and the mission would continue as an earth orbit alternate, or if other conditions warranted, to landing in the West Atlantic or Central Pacific after one revolution. Mode IV is preferred over Mode III. A variation of Mode IV is the Apogee Kick in which the SPS would be ignited at first apogee to raise perigee and thereby set up a suitable orbit for a low earth-orbit alternate mission.

Earth Parking Orbit phase--

Aborts from earth parking orbit would be flown similar to the normal deorbit and entry that was flown on Apollo 7: SPS deorbit burn followed by CM/SM separation and guided entry.

Translunar Injection Phase --

Aborts during the translunar injection phase are only
a remote possibility, but if an abort became necessary
during the TLI maneuver, an SPS retrograde burn could be
made to produce spacecraft entry. This mode of abort would
be used only in the event of an extreme emergency that
affected crew safety. The spacecraft landing point would
vary with launch azimuth and length of the TLI burn. Another
TLI abort situation would be used if a malfunction cropped
up after injection. A retrograde SPS burn at about 90
minutes after TLI shutoff would allow targeting to land in
an Atlantic contigency landing area between $7\frac{1}{2}$ and $17\frac{1}{2}$ hours
after initiating abort, depending on the change in velocity
applied.

Translunar Coast phase--

Aborts arising during the three-day translunar coast
phase would be similar in nature to the 90-minute TLI abort.
Aborts from deep space bring into the play the moon's antipode (line projected from moon's center through earth's
center to opposite face) and the effect of the earth's rotation upon the geographical location of the antipode. Abort
times would be selected for landing when the antipode crosses
165 WLong. The antipode crosses the mid-Pacific recovery
line once each 24 hours, and if a time-critical situation
forces an abort earlier than the selected fixed abort times,
landings would be targeted for the Atlantic Ocean, East
Pacific, West Pacific or Indian Ocean recovery lines in that
order of preference. From TLI plus 44 hours, a circumlunar
abort becomes faster than an attempt to return directly to
earth.

Lunar Orbit Insertion phase--

Early SPS shutdowns during the lunar orbit insertion
burn (LOI) are covered by Modes I and III in the Apollo 8
mission (Mode II involves lunar module operations). Both
modes would result in the CM landing the earth latitude of
the moon antipode at the time the abort was performed.
Mode I would be an SPS posigrade burn into an earth-return
trajectory as soon as possible following LOI shutdown during
the first two minutes of the LOI burn. Mode III occurs near
pericynthion following one or more revolutions in lunar orbit.
Following one or two lunar orbits, the Mode III posigrade SPS
burn at pericynthion would inject the spacecraft into a transearth trajectory targeted for the mid-Pacific recovery line.

- more -

Lunar Orbit Phase --

If during lunar parking orbit it became necessary to abort, the transearth injection (TEI) burn would be made early and would target spacecraft landing to the mid-Pacific recovery line.

Transearth Injection phase--

Early shutdown of the TEI burn between ignition and two minutes would cause a Mode III abort and a SPS posigrade TEI burn would be made at a later pericynthion. Cutoffs after two minutes TEI burn time would call for a Mode I abort---restart of SPS as soon as possible for earth-return trajectory. Both modes produce mid-Pacific recovery line landings near the latitude of the antipode at the time of the TEI burn.

Transearth Coast phase--

Adjustments of the landing point are possible during the transearth coast through burns with the SPS or the service module RCS thrusters, but in general, these are covered in the discussion of transearth midcourse corrections. No abort burns will be made later than 20 hours prior to entry to avoid effects upon CM entry velocity and flight path angle.

- more -

PHOTOGRAPHIC TASKS

Photography seldom before has played as important a role in a spaceflight mission as it will on Apollo 8. The crew will have the task of photographing not only a lunar surface Apollo landing site to gather valuable data for subsequent lunar landing missions, but will also point their cameras toward visual phenomena in cislunar space which heretofore have posed unanswered questions.

A large quantity of film of various types has been loaded aboard the Apollo 8 spacecraft for lunar surface photography and for items of interest that crop up in the course of the mission.

Camera equipment carried on Apollo 8 consists of two 70mm Hasselblad still cameras with two 80mm focal length lenses, a 250mm telephoto lens, and associated equipment such as filters, ringsight, spotmeter and intervalometer for stereo strip photography. For motion pictures a 16mm Maurer data acquisition camera with variable frame speed selection will be used. Accessories for the motion picture camera include lenses of 200, 75, 18 and 5mm focal lengths, a right-angle mirror, a command module boresight bracket and a power cable.

Photographic tasks have been divided into three general categories: lunar stereo strip photography, engineering photography and items of interest.

Apollo 8 photographic tasks are summarized as follows:

Lunar Stereo Strip Photography -- Overlapping stereo 70mm frames shot along the lunar orbit ground track with spacecraft aligned to local vertical. Photos will be used for terrain analysis and photometric investigations.

Engineering Photography -- Through-the-window photography of immediate region around spacecraft to gather data on existence of contaminant cloud around the spacecraft and to further understand source of window visibility degradation. Cabin interior photography documenting crew activities will also be taken as an aid to following flight crews.

Items of Interest -- Dim-light targets: Gegenschein (a round or elongated spot of light in space at a point 180° from the sun) photos on one-minute exposure with spacecraft held in inertial attitude on dark side on Moon and during translunar and transearth coast; Zodiacal light along the plane of the ecliptic (path of Sun around celestial sphere), one-minute exposures during dark side of lunar orbit; Star fields under various lighting conditions to study effect of spacecraft debris clouds and window contamination on ability to photograph stars; lunar surface in earthshine to gain photometric data about lunar surface under low-level illumination.

Lunar surface in daylight at zero phase angle (spacecraft shadow directly below spacecraft) to further measure reflective properties of lunar surface; lunar terminator in daylight at oblique angles to evaluate capability of terrain analysis on such photos; Apollo exploration sites, Surveyor landing sites and specific features and areas to augment present lunar surface photography and to correlate with Surveyor photos; image motion compensation with long-length lenses by tracking target with spacecraft; phenomena, features and other items of interest selected by the crew in real time; and lunar seas through red and blue filters for correlating with color discontinuities observed from earth.

Apollo 8 film stowage is as follows: 3 magazines of Panatomic-X intermediate speed black and white for total 600 frames; 2 magazines SO-368 Ektachrome color reversal for total 352 frames; 1 magazine SO-121 Ektachrome special daylight color reversal for total 160 frames; and 1 magazine 2485 high-speed black and white (ASA 6,000, push to 16,000) for dim-light photography, total 120 frames. Motion picture film: nine 130-foot magazines SO-368 for total 1170 feet, and two magazines SO-168 high speed interior color for total 260 feet.

TELEVISION

As many as seven live television transmissions are being considered during the Apollo 8 flight. Up to three transmissions are being considered on the way to the Moon, one to two from lunar orbit, and possibly two on the way back from the Moon.

The frequency and duration of TV transmissions are dependent upon the level of various other mission activities and the availability of the spacecraft's high gain antenna. This antenna is primarily used to transmit engineering data, which has priority over TV transmissions.

The TV signals will be sent from the spacecraft to ground stations at Goldstone, California and Madrid, Spain, where the signal will be converted to commercial frequencies. The TV signal will be released live to public networks from the Mission Control Center, Houston.

Because of crew activities from launch through translunar injection, TV operations are not planned prior to the translunar coast phase of the mission about 12 hours into the flight.

The purpose of television during Apollo 8 is to evaluate TV transmission at lunar distances for planning future lunar missions and to provide live TV coverage of the Apollo 8 flight to the public.

The 4.5 pound RCA TV camera is equipped with 160 degree and 9 degree field of view lens. A 12-foot power-video cable permits the camera to be hand-held at the command module windows for the planned photography.

In TV broadcast to homes, the average distance is only five miles between the transmitting station and the home TV set, while the station transmits an average of 50,000 watts of power. In contrast, the Apollo 8 TV camera operates on only 20 watts, and on this mission, will be over 200,000 miles from the home TV sets.

The NASA ground station's large antenna and sensitive receivers make up for most of this difference, but the Apollo pictures are not expected to be as high quality as normal broadcast programs.

SPACECRAFT STRUCTURE SYSTEMS

Apollo spacecraft No. 103 for the Apollo 8 mission is comprised of a launch escape system, command module, service module and a spacecraft-lunar module adapter. The latter serves as a mating structure to the instrument unit atop the S-IVB stage of the Saturn V for this mission, lunar module test article B (LTA-B) will be housed in the adapter.

Launch Escape System/(LES)--Propels command module to safety in an aborted launch. It is made up of an open-frame tower structure mounted to the command module by four frangible bolts, and three solid-propellant rocket motors: a 155,000-pound-thrust launch escape system motor, a 3,000-pound-thrust pitch control motor that bends the command module trajectory away from the launch vehicle and pad area. Two canard vanes near the top deploy to turn the command module aerodynamically to an attitude with the heat-shield forward. Attached to the base of the Escape System is a boost protective cover composed of glass, cloth and honeycomb, that protects the command module from rocket exhaust gases from the main and the jettison motor. The system is 33 feet tall, four feet in diameter at the base and weighs 8,900 pounds (4040 kg).

Command Module (CM) Structure--The basic structure of the command module is a pressure vessel encased in heat-shields, cone-shaped 12 feet high, base diameter of 12 feet 10 inches, and launch weight 12,392 pounds (5626 kg).

The command module consists of the forward compartment which contains two negative pitch reaction control engines and components of the Earth landing system; the crew compartment, or inner pressure vessel, containing crew accommodations, controls and displays, and spacecraft systems; and the aft compartment housing ten reaction control engines and fuel tankage.

Heat-shields around the three compartments are made of brazed stainless steel honeycomb with an outer layer of phenolic epoxy resin as an ablative material. Heat-shield thickness, varying according to heat loads, ranges from 0.7 inches (at the apex) to 2.7 inches on the aft side.

The spacecraft inner structure is of aluminum alloy sheet-aluminum honeycomb bonded sandwich ranging in thickness from 0.25 inches thick at forward access tunnel to 1.5 inches thick at base.

Service Module (SM) Structure--The service module is a
cylinder 12 feet 10 inches in diameter by 22 feet long. For
the Apollo 8 mission, it will weigh 51,258 pounds (23,271 kg)
at launch. Aluminum honeycomb panels one inch thick form the
outer skin, and milled aluminum radial beams separate the interior into six sections containing service propulsion system and
reaction control fuel-oxidizer tankage, fuel cells and onboard
consumables.

Spacecraft-LM Adapter (SLA) Structure--The spacecraft-
LM adapter is a truncated cone 28 feet long tapering from 260
inches diameter at the base to 154 inches at the forward end
at the service module mating line. Aluminum honeycomb 1.75
inches thick is the stressed-skin structure for the spacecraft
adapter. The SLA weighs 4,150 pounds (1,884 kg).

Spacecraft Systems

Guidance, Navigation and Control System/(GNCS)--Measures
and controls spacecraft attitude and velocity, calculates
trajectory, controls spacecraft propulsion system thrust vector
and displays abort data. The Guidance System consists of
three subsystems: inertial, made up of inertial measuring unit
and associated power and data components; computer, consisting
of display and keyboard panels and digital computer which processes information to or from other components; and optic, including scanning telescope, sextant for celestial and/or landmark spacecraft navigation.

Stabilization and Control System/(SCS)--Controls spacecraft rotation, translation and thrust vector and provides
displays for crew-initiated maneuvers; backs up the guidance
system. It has three subsystems; attitude reference, attitude
control and thrust vector control.

Service Propulsion System/(SPS)--Provides thrust for
large spacecraft velocity changes and de-orbit burn through
a gimbal-mounted 20,500-pound-thrust hypergolic engine using
nitrogen tetroxide oxidizer and a 50-50 mixture of unsymmetrical
dimethyl hydrazine and hydrazine fuel. Tankage of this system
is in the service module. The system responds to automatic
firing commands from the guidance and navigation system or to
manual commands from the crew. The engine provides a constant
thrust rate. The stabilization and control system gimbals the
engine to fire through the spacecraft center of gravity.

Reaction Control System/(RCS)--This includes two
independent systems for the command module and the service
module. The service module reaction controls have four
identical quads of four 100-pound thrust hypergolic engines
mounted, near the top of the Service Module, 90 degrees apart
to provide redundant spacecraft attitude control through cross-
coupling logic inputs from the Stabilization and Guidance
Systems. Small velocity change maneuvers can also be made
with the Service Module reaction controls. The Command Module
Reaction Control System consists of two independent six-engine
subsystems of 94 pounds thrust each. One is activated after
separation from the Service Module, and is used for spacecraft
attitude control during entry. The other is maintained in a
sealed condition as a backup. Propellants for both systems
are monomethyl hydrazine fuel and nitrogen tetroxide oxidizer
with helium pressurization. These propellants are hypergolic,
i.e.: they burn spontaneously on contact without need for an
igniter.

Electrical Power System/(EPS)--Consists of three, 31-
cell Bacon-type hydrogen-oxygen fuel cell power plants in the
Service Module which supply 28-volt DC power, three 28-volt
DC zinc-silver oxide main storage batteries in the Command
Module lower equipment bay, two pyrotechnic batteries in the
Command Module lower equipment bay, and three 115-200-volt
400-cycle three-phase AC inverters powered by the main 28-volt
DC bus. The inverters are also located in the lower equipment
bay. Supercritical cryogenic hydrogen and oxygen react in the
fuel cell stacks to provide electrical power, potable water and
heat. The Command Module main batteries can be switched to fire
pyrotechnics in an emergency. A battery charger builds the
batteries to full strength as required.

Environmental Control System/(ECS)--Controls spacecraft
atmosphere, pressure and temperature and manages water. In
addition to regulating cabin and suit gas pressure, temperature
and humidity, the system removes carbon dioxide, odors and
particles, and ventilates the cabin after landing. It collects
and stores fuel cell potable water for crew use, supplies water
to the glycol evaporators for cooling, and dumps surplus water
overboard through the urine dump valve. Excess heat generated
by spacecraft equipment and crew is routed by this system to
the cabin heat exchangers, to the space radiators, to the glycol
evaporators, or it vents the heat to space.

Telecommunication System--Consists of pulse code modulated telemetry for relaying to Manned Space Flight Network stations data on spacecraft systems and crew condition, VHF/AM and unified S-Band tracking transponder, air-to-ground voice communications, onboard television, and a VHF recovery beacon. Network stations can transmit to the spacecraft such items as updates to the Apollo guidance computer and central timing equipment, and real-time commands for certain onboard functions.

The Apollo high-gain steerable S-Band antenna will be flown for the first time on the Apollo 8 mission. Deployed shortly after CSM separation from the S-IVB stage, the high-gain antenna will be tested in the two-way mode between the spacecraft and the Manned Space Flight Network stations during translunar coast, lunar orbit and Earth return.

The high-gain S-Band antenna consists of four, 31-inch-diameter parabolic dishes mounted on a folding boom at the aft end of the service module. Nested alongside the service propulsion system engine nozzle until deployment, the antenna swings out at right angles to the spacecraft longitudinal axis, with the boom pointing 52 degrees below the heads-up horizontal. Signals from the ground stations can be tracked either automatically or manually with the antenna's gimballing system. All normal S-Band voice and uplink/downlink communications will be handled by the high-gain antenna.

Sequential System--Interfaces with other spacecraft systems and subsystems to initiate critical functions during launch, docking maneuvers, pre-orbital aborts and entry portions of a mission. The system also controls routine spacecraft sequencing such as Service Module separation and deployment of the Earth landing system.

Emergency Detection System/(EDS)--Detects and displays to the crew launch vehicle emergency conditions, such as excessive pitch rates or two engines out, and automatically or manually shuts down the booster and activates the launch escape system; functions until the spacecraft is in orbit.

Earth Landing System/(ELS)--Includes the drogue and main parachute system as well as post-landing recovery aids. In a normal entry descent, the Command Module apex cover is jettisoned at 24,000 feet, followed by two mortar-deployed reefed 16.5-foot diameter drogue parachutes for orienting and decelerating the spacecraft. After drogue release, three pilot chutes pull out the three main 83.3-foot diameter parachutes with two-stage reefing to provide gradual inflation in three steps. Two main parachutes out of three will provide a safe landing.

Recovery aids include the uprighting system, swimmer interphone connections, sea dye marker, flashing beacon, VHF recovery beacon and VHF transceiver. The uprighting system consists of three compressor-inflated bags to turn the spacecraft upright if it should land in the water apex down (Stable II position).

Caution and Warning System--Monitors spacecraft systems for out-of-tolerance conditions and alerts crew by visual and audible alarms so that crewmen may trouble-shoot the problem.

Controls and Displays--Provide readouts and control functions of all other spacecraft systems in the command and service modules. All controls are designed to be operated by crewmen in pressurized suits. Displays are grouped according to the frequency the crew refers to them.

Q-BALL (NOSE CONE)

PITCH CONTROL MOTOR

CANARDS

JETTISON MOTOR

LAUNCH ESCAPE MOTOR

STRUCTURAL SKIRT

LAUNCH ESCAPE TOWER

TOWER ATTACHMENT (4)

COMMAND MODULE

BOOST PROTECTIVE COVER

EPS RADIATOR

REACTION CONTROL SYSTEM ENGINES

SERVICE MODULE

ECS RADIATOR

SPS ENGINE EXPANSION NOZZLE

SPACECRAFT LM ADAPTER (SLA)

SLA PANEL JUNCTION (BETWEEN FWD AND AFT PANELS)

NOTE: Lunar Module Test Article will be flown on this mission instead of Lunar Module.

INSTRUMENT UNIT (SHOWN AS REFERENCE)

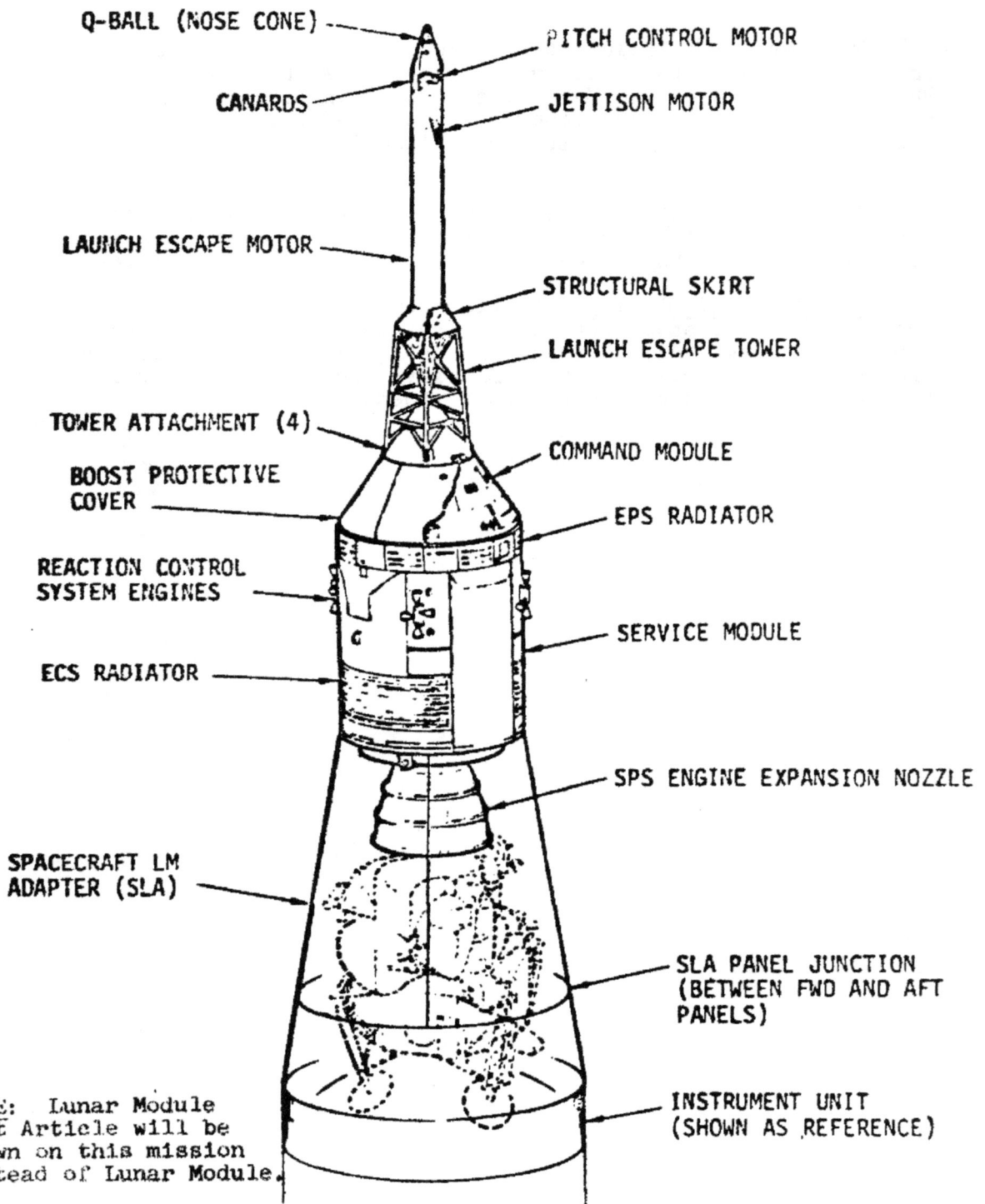

SPACECRAFT CONFIGURATION

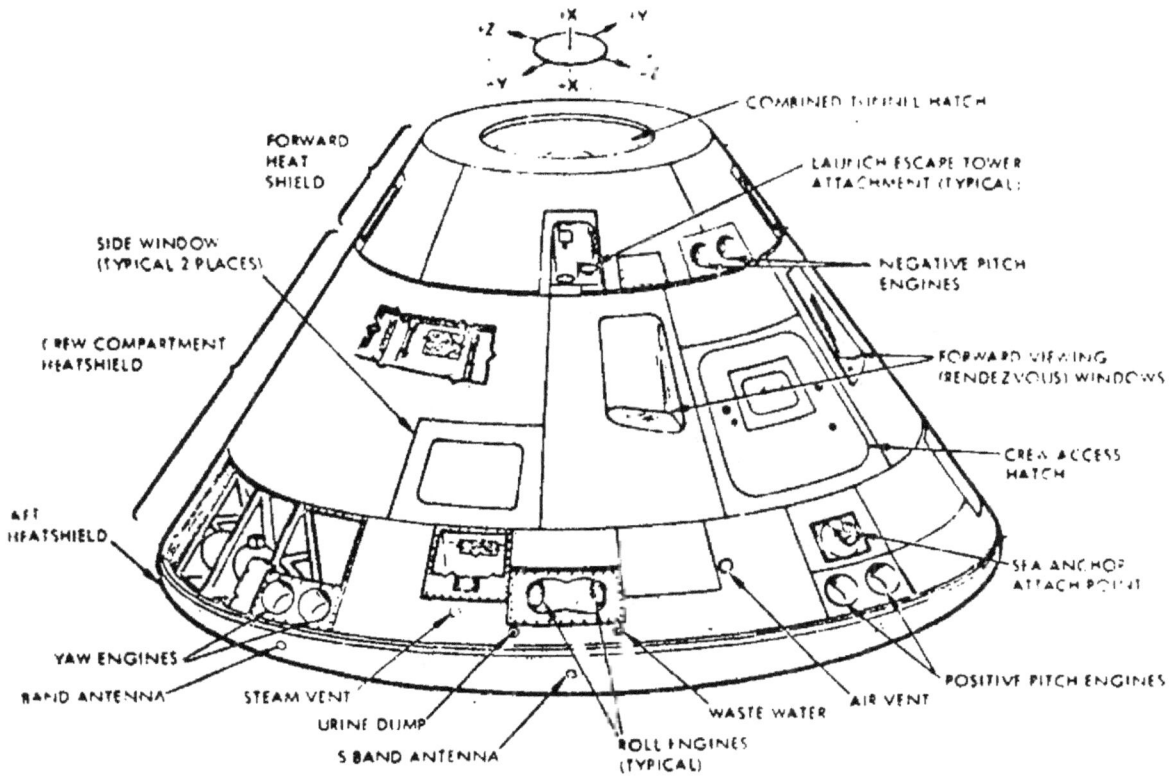

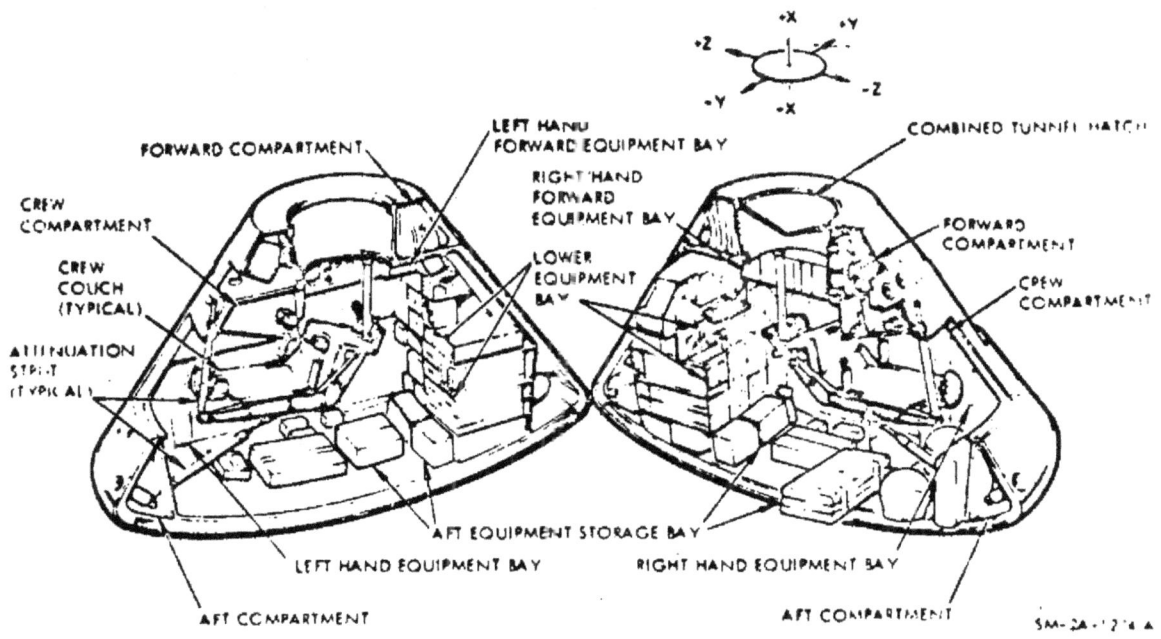

FORWARD HEAT SHIELD

SIDE WINDOW (TYPICAL 2 PLACES)

CREW COMPARTMENT HEATSHIELD

AFT HEATSHIELD

YAW ENGINES

BAND ANTENNA

STEAM VENT

URINE DUMP

S BAND ANTENNA

ROLL ENGINES (TYPICAL)

WASTE WATER

AIR VENT

POSITIVE PITCH ENGINES

SEA ANCHOR ATTACH POINT

CREW ACCESS HATCH

FORWARD VIEWING (RENDEZVOUS) WINDOWS

NEGATIVE PITCH ENGINES

LAUNCH ESCAPE TOWER ATTACHMENT (TYPICAL)

COMBINED TUNNEL HATCH

FORWARD COMPARTMENT

CREW COMPARTMENT

CREW COUCH (TYPICAL)

ATTENUATION STRUT (TYPICAL)

AFT COMPARTMENT

LEFT HAND EQUIPMENT BAY

AFT EQUIPMENT STORAGE BAY

LEFT HAND FORWARD EQUIPMENT BAY

RIGHT HAND FORWARD EQUIPMENT BAY

LOWER EQUIPMENT BAY

RIGHT HAND EQUIPMENT BAY

AFT COMPARTMENT

COMBINED TUNNEL HATCH

FORWARD COMPARTMENT

CREW COMPARTMENT

SM-2A-1274A

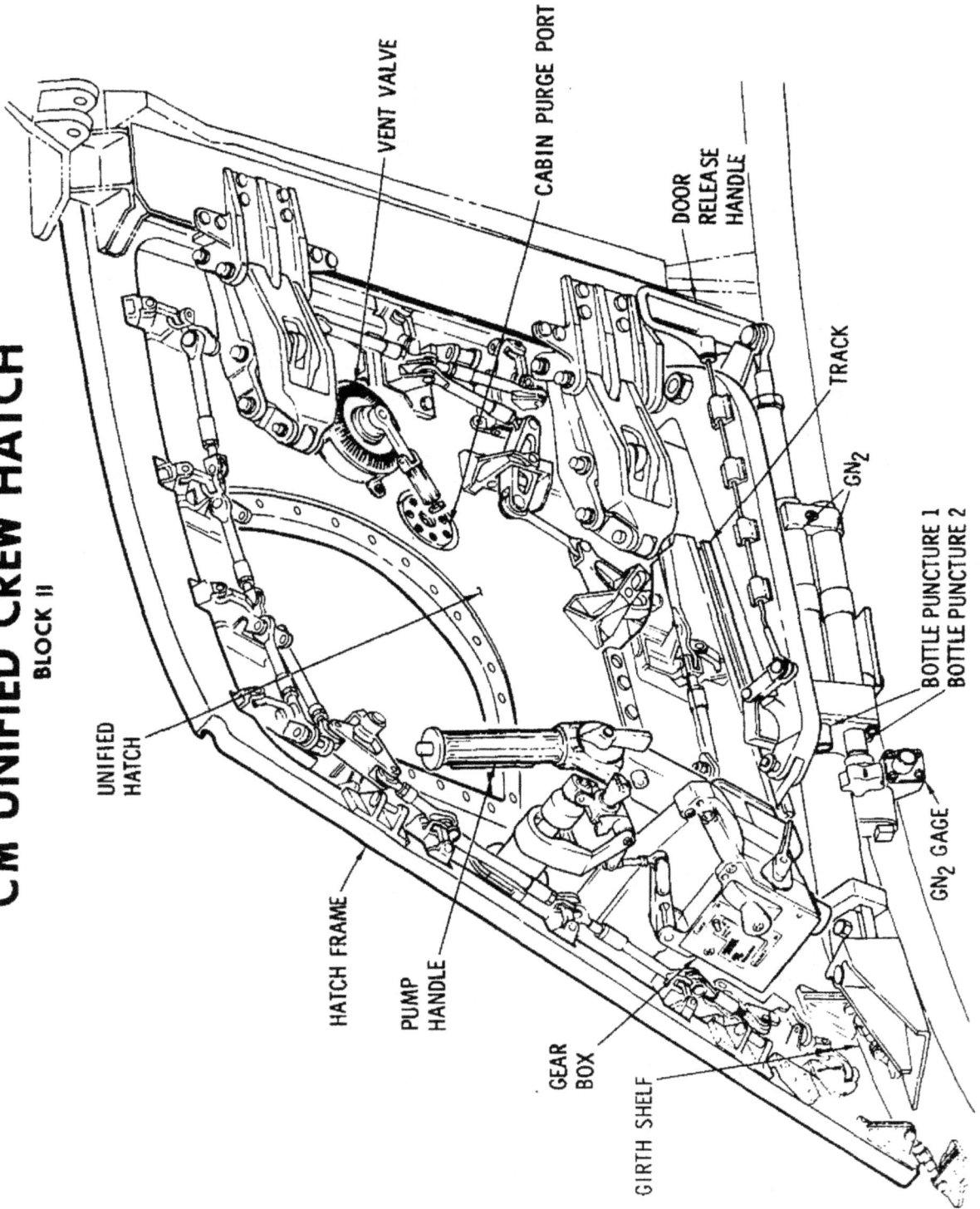

CM UNIFIED CREW HATCH

BLOCK II

VENT VALVE

CABIN PURGE PORT

DOOR RELEASE HANDLE

TRACK

GN$_2$

BOTTLE PUNCTURE 1
BOTTLE PUNCTURE 2

GN$_2$ GAGE

UNIFIED HATCH

HATCH FRAME

PUMP HANDLE

GEAR BOX

GIRTH SHELF

SERVICE MODULE
BLOCK II

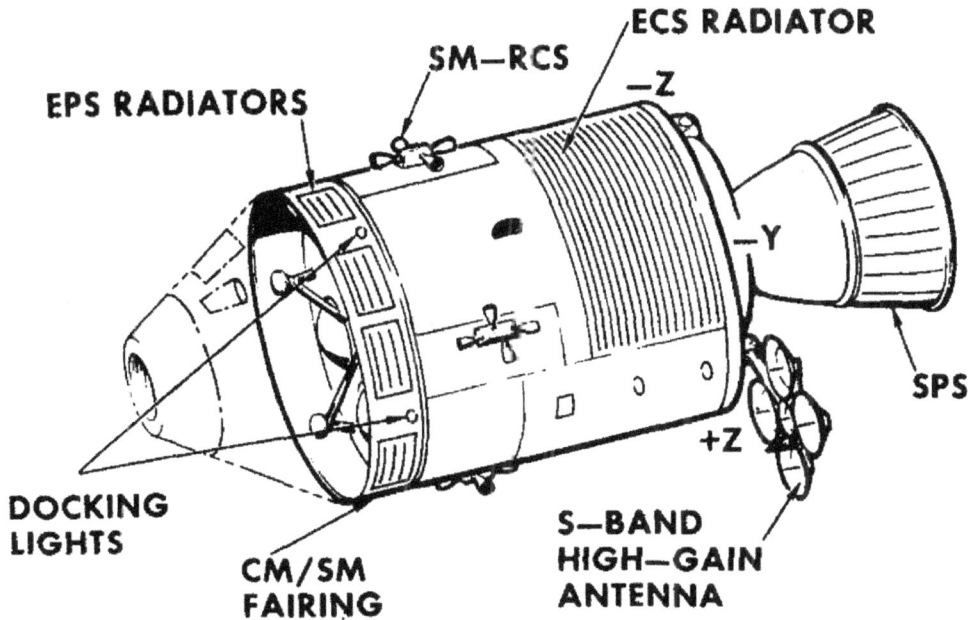

EPS RADIATORS

SM—RCS

ECS RADIATOR

—Z

—Y

SPS

DOCKING LIGHTS

CM/SM FAIRING

+Z

S—BAND HIGH—GAIN ANTENNA

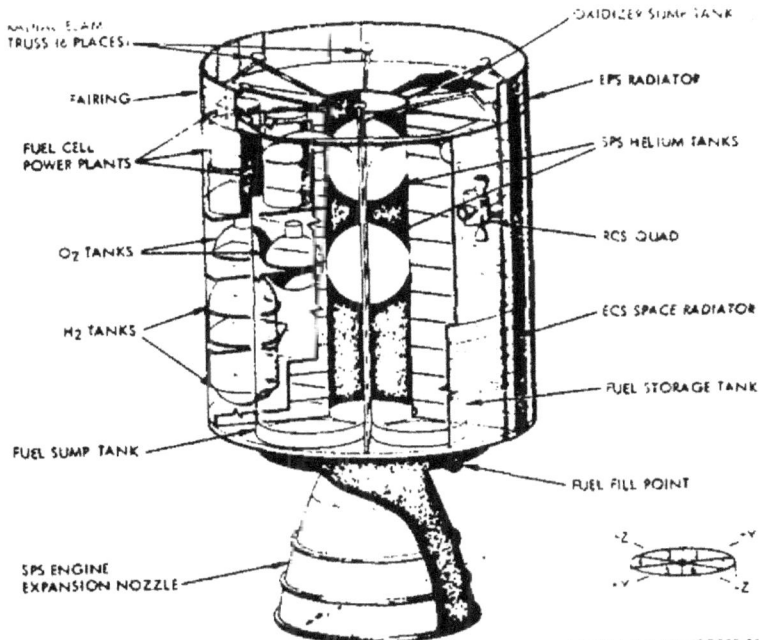

RADIAL BEAM TRUSS (6 PLACES)

OXIDIZER SUMP TANK

FAIRING

EPS RADIATOR

FUEL CELL POWER PLANTS

SPS HELIUM TANKS

O₂ TANKS

RCS QUAD

H₂ TANKS

ECS SPACE RADIATOR

FUEL STORAGE TANK

FUEL SUMP TANK

FUEL FILL POINT

SPS ENGINE EXPANSION NOZZLE

1 AND 4 ARE 50-DEGREE SECTORS
2 AND 5 ARE 70-DEGREE SECTORS
3 AND 6 ARE 60-DEGREE SECTORS

SATURN V LAUNCH VEHICLE

The white Saturn V launch vehicle, with the Apollo spacecraft and launch escape system mounted atop, towers 363 feet above the launch pad. The three propulsive stages and the instrument unit have a combined height of 281 feet. The vehicle weighs 6,219,760 pounds at ignition.

Marked with black paint in sections for better optical tracking, identified with huge red lettering, and wearing the United States Flag on the first stage, the giant vehicle is capable of hurling 285,000 pounds into low Earth orbit or sending about 100,000 pounds to the Moon.

First Stage

The first stage (S-IC) of the Saturn V is 138 feet tall and 33 feet in diameter, not including the fins and engine shrouds on the thrust structure. It was developed jointly by the National Aeronautics and Space Administration's Marshall Space Flight Center, Huntsville, Ala., and The Boeing Co.

Marshall assembled four S-IC stages: a structural test model, a static test version and the first two flight stages. The first flight stage launched Apollo 4 on the first Saturn V flight Nov. 9, 1967. The second S-IC launched Apollo 6 on April 4, 1968.

Boeing, as prime contractor, built two ground test units. Boeing is responsible for assembly of the other 13 flight stages at Marshall's Michoud Assembly Facility in New Orleans. The first flight model S-IC built by Boeing is the first stage of the AS-503 launch vehicle.

The static test model and the first three flight versions were fired at the Marshall Space Flight Center Test Laboratory. All other S-IC stages are being test fired at Marshall's Mississippi Test Facility in Hancock County, Miss.

Dry weight of the first stage is 305,650 pounds. Its two propellant tanks have a total capacity of 4.4 million pounds of fuel and oxidizer -- some 202,000 gallons (1,352,711 pounds) of RP-1 (kerosene) and 329,000 gallons (3,130,553 pounds) of liquid oxygen. Stage weight at separation, including residual propellants, will be 380,738 pounds. The normal propellant flow rate to the five F-1 engines is 28,000 pounds per second. The five engines produce a combined thrust roughly equivalent to 180 million horsepower at maximum speed.

-more-

During the planned 151 seconds of burn time, the engines will propel the Apollo/Saturn V to an altitude of 36.3 nautical miles (41.9 sm, 67 km) and carry it downrange 47.4 nautical miles (54.6 sm, 88 km), making good a speed of 5,267.3 knots (6,068 statute miles-per-hour) at first stage engine cutoff.

Four of the engines are mounted on a ring, each 90 degrees from its neighbor. These four can be gimballed to control the rocket's direction of flight.

The fifth engine is mounted rigidly in the center.

Second Stage

The second stage (S-II) is 81.5 feet tall and 33 feet in diameter. It weighs 88,600 pounds dry, 1,035,463 pounds loaded with propellant. Weight at separation will be 103,374 pounds.

The 14,774 pounds difference between dry weight and weight at separation includes the 12,610-pound S-IC/S-II interstage section, 2,164 pounds of ullage rocket propellants and other items on board.

The stage's two propellant tanks carry about 271,800 gallons (152,638 pounds) of liquid hydrogen and 87,500 gallons (792,714 pounds) of liquid oxygen. Its five J-2 engines develop a combined thrust of 1 million pounds.

The second stage carries the rocket to an altitude of 105.8 nautical miles (121.9 sm, 197 km) and a distance of some 805 nautical miles (927.4 sm, 1490 km) downrange. Before burnout it will be moving 13,245 knots (15,258.3 mph). The J-2 engines will run six minutes and seven seconds.

The Space Division of North American Rockwell Corp., builds the second stage at Seal Beach, Calif. The cylindrical vehicle is made up of the forward skirt (to which the third stage connects), the liquid hydrogen tank, the liquid oxygen tank, the thrust structure (on which the engines are mounted) and an interstage section (to which the first stage connects). The tanks are separated by an insulated common bulkhead.

North American Rockwell conducted research and development static testing at the Santa Susana, Calif., test facility and at the NASA-Mississippi Test Facility. The flight stage for the Apollo 8 was shipped via the Panama Canal for captive firings at Mississippi Test Facility.

Third Stage

The third stage (S-IVB) was developed by the McDonnell Douglas Astronautics Co. at Huntington Beach, Calif. It is the larger and more powerful successor to the S-IV that served as the second stage of the Saturn I.

The third stage is flown from its manufacturing site to the McDonnell Douglas' Test Center, Sacramento, Calif., for static test firings. The stage is then flown to the NASA-Kennedy Space Center.

Measuring 58 feet 5 inches long by 21 feet 8 inches in diameter, the stage weighs 26,000 pounds dry. At separation in flight its weight will be 29,754 pounds exclusive of the liquid hydrogen and liquid oxygen in the main tanks. This extra weight consists mainly of solid and liquid propellants used in retro and ullage rockets and in the auxiliary propulsion system (APS).

An interstage section connects the second and third stages. This 8,760-pound section stays with the second stage at separation, exposing the single J-2 engine mounted on the thrust structure. The after skirts connected to the interstage at the separation plane, encloses the liquid oxygen tank which holds some 20,400 gallons of the oxidizer. Above this is the large fuel tank holding about 77,200 gallons of liquid hydrogen. Weight of the S-IVB and payload at insertion into parking orbit will be 283,213 pounds. Weight at injection into translunar trajectory will be 122,380 pounds.

Total usable propellants carried in the two tanks is 234,509 pounds, with fuel and oxidizer separated by an insulated common bulkhead. Insulation is necessary in both upper stages because liquid oxygen, at about 293 degrees below zero F, is too warm for liquid hydrogen, at minus 423 degrees.

The aft skirt also serves as a mount for two auxiliary propulsion system modules spaced 180 degrees apart. Each module contains three liquid-fueled 147-pound thrust engines, one each for roll, pitch and yaw, and a 72-pound-thrust, liquid-fueled ullage engine.

Four solid-propellant retro-rockets of 37,500 pounds thrust each are mounted on the interstage to back the second stage away from the third stage at separation. The third stage also carried two solid-propellant ullage motors of 3,400 pounds thrust each. These motors help to move the third stage forward and away from the second stage upon separation and serve the additional purpose of settling the liquid propellants in the bottoms of the tanks in preparation for J-2 ignition. The first J-2 burn is 152 seconds, the second, 5 min. 12 sec.

Propulsion

The 41-rocket engines of the Saturn V have thrust ratings ranging from 72 pounds to more than 1.5 million pounds. Some engines burn liquid propellants, others use solids.

The five F-1 engines in the first stage burn RP-1 (kerosene) and liquid oxygen. Each engine in the first stage develops 1.415 million pounds of thrust at liftoff, building up to 1.7 million pounds thrust before cutoff. The cluster of five F-1s gives the first stage a thrust range from 7.57 million pounds at liftoff to 8.5 million pounds just before cutoff.

The F-1 engine weighs almost 10 tons, is more than 18 feet high and has a nozzle-exit diameter of nearly 14 feet. The F-1 undergoes static testing for an average 650 seconds in qualifying for the 150-second run during the Saturn V first stage booster phase. This run period, 800 seconds, is still far less than the 2,200 seconds of the engine guarantee period. The engine consumes almost three tons of propellants per second.

The first stage of the Saturn V for this mission has four other rocket motors. These are the solid-fuel retro-rockets which will slow and separate the stage from the second stage. Each rocket produces a thrust of 87,900 pounds for 0.6 second.

The main propulsion for the second stage is a cluster of five J-2 engines burning liquid hydrogen and liquid oxygen. Each engine develops a mean thrust of 200,000 pounds (variable from 175,000 to 225,000 in phases of flight), giving the stage a total mean thrust of 1 million pounds.

Designed to operate in the hard vacuum of space, the 3,500-pound J-2 is more efficient than the F-1 because it burns the high-energy fuel hydrogen.

The second stage also has four 21,000-pound-thrust solid-fuel rocket engines. These are the ullage rockets mounted on the interstage section. These rockets fire to settle liquid propellant in the bottom of the main tanks and help attain a "clean" separation from the first stage, then they drop away with the interstage at second plane separation.

-more-

Fifteen rocket engines perform various functions on the third stage. A single J-2 provides the main propulsive force; there are two main ullage rockets, four retro-rockets and eight smaller engines in the auxiliary propulsion system.

Instrument Unit

The Instrument Unit (IU) is a cylinder three feet high and 21 feet 8 inches in diameter. It weighs 4,880 pounds.

Components making up the "brain" of the Saturn V are mounted on cooling panels fastened to the inside surface of the instrument unit skin. The refrigerated "cold plates" are part of a system that removes heat by circulating fluid coolant through a heat exchanger that evaporates water from a separate supply into the vacuum of space.

The six major systems of the instrument unit are structural, thermal control, guidance and control, measuring and telemetry, radio frequency and electrical.

The instrument unit maintains navigation, guidance and control of the vehicle; measurement of vehicle performance and environment; data transmission with ground stations; radio tracking of the vehicle; checkout and monitoring of vehicle functions; detection of emergency situations; generation and network distribution of electric power for system operation; and preflight checkout and launch and flight operations.

A path-adaptive guidance scheme is used in the Saturn V instrument unit. A programmed trajectory is used in the initial launch phase with guidance beginning only after the vehicle has left the atmosphere. This is to prevent movements that might cause the vehicle to break apart while attempting to compensate for winds, jet streams and gusts encountered in the atmosphere.

If such air currents displace the vehicle from the optimum trajectory in climb, the vehicle derives a new trajectory. Calculations are made about once each second throughout the flight. The launch vehicle digital computer and launch vehicle data adapter perform the navigation and guidance computations.

The ST-124M inertial platform -- the heart of the navigation, guidance and control system -- provides space-fixed reference coordinates and measures acceleration along the three mutually perpendicular axes of the coordinate system.

International Business Machines Corp. is prime contractor for the instrument unit and is the supplier of the guidance signal processor and guidance computer. Major suppliers of instrument unit components are: Electronic Communications, Inc., control computer; Bendix Corp., ST-124M inertial platform; and IBM Federal Systems Division, launch vehicle digital computer and launch vehicle data adapter.

Launch Vehicle Camera Systems

Fewer cameras will be carried aboard the Saturn V launch vehicle on the Apollo 8 mission than on the previous Saturn V flights. The first stage will carry four motion picture cameras and two television cameras. The film cameras will be ejected for recovery.

First Stage -- Film Cameras

The four film cameras will be mounted on the inside of the forward skirt of the first stage. Two cameras (1 and 3) will be mounted lens forward and canted inward five degrees to view the separation of the first and second stages. These two cameras will start 144 seconds after liftoff and run about 40 seconds. Two cameras (2 and 4) will be mounted lens-aft, with lenses connected by fiber optic bundles to manhole covers in the top of the liquid oxygen tank. These covers provide viewing windows and mounts for strobe lights. The tank will be lighted inside by pulsed strobe to enable the cameras to record the behavior of the liquid oxygen in flight. These two cameras will be turned on 30 seconds before liftoff; the strobe lights will be turned off shortly before first stage cutoff.

The film cameras, loaded with color film, are carried in recoverable capsules inserted in ejection tubes. They will be ejected 177 seconds after vehicle liftoff, or 25 seconds after stage separation, at 49 nautical miles (56.4 sm, 90.8 km) altitude at a point 76.8 nautical miles (88.5 sm, 142 km) downrange. Camera impact is expected some 413.2 nm (473 sm, 763 km) downrange about 11 minutes after liftoff.

First Stage -- Television

Both television cameras will be mounted inside the thrust structure of the first stage.

A fiber optic bundle from each camera will split into two separate bundles going to lenses mounted outside the heatshield in the engine area. This will provide two images for each camera, or four for the system. The images will be tilted 90 degrees from vertical to give a wider view as two images appear on each cathode ray tube. Each lens will view the center engine and one outer engine, thus providing a view of each outer engine working in conjunction with the fixed center engine.

A removable aperture disk can be changed from f/2 to f/22 to vary the image intensity. (Fiber optics reduce image intensity about 70 per cent.) A quartz window, rotated by a DC motor with a friction drive, protects the objective lens on the end of the fiber optic bundle. Fixed metallic mesh scrapers will remove soot from the rotating windows. Images from the two objective lenses are combined into the dual image in the larger fiber optics bundle by a "T" fitting. A 14-element coupling lens adapts the large dual image bundle to the camera.

Images from both cameras are multiplexed together for transmission on a single telemetry link. The images are unscrambled at the receiving station.

The video system cameras contain 28-volt vidicon cameras, pre-amplifiers and vertical sweep circuits for 30 frames per second scanning.

The TV cameras operate continuously from 55 minutes before liftoff until destroyed on first stage reentry.

Ejection and Recovery

When the camera capsules are ejected, stabilization flaps open for the initial part of the descent. When the capsules descend to about 15,000 feet above ground a para-balloon will inflate automatically, causing flaps to fall away. A recovery radio transmitter and flashing light beacon are turned on about 6 seconds after para-balloon inflation. After touchdown, the capsule effuses a dye marker to aid sighting, and it releases a shark repellent to protect the capsules, para-balloon (which keeps the capsule afloat) and the recovery team.

A Navy ship with helicopters and frogmen aboard will be cruising in the splashdown area for the camera capsules. The capsules will be picked up and flown by helicopter to Kennedy Space Center. The capsules will then be transferred to a data plane at Patrick Air Force Base, Fla., for immediate transfer to the Marshall Center at Huntsville where the film will be processed.

Sequence of Events

NOTE: Information presented in this press kit is
based on a nominal mission. Plans may be altered
prior to or during flight to meet changing conditions.

Launch

The first stage of the Saturn V will carry the vehicle
and Apollo spacecraft to an altitude of 36.3 nautical miles
(41.9 sm, 67 km) and 47.4 nautical miles (54.6 sm, 88 km)
downrange, building up speed to 5,267.3 knots (6,068 mph)
in two minutes 31 seconds of powered flight.
After separation from the second stage, the first stage
will continue a ballistic trajectory ending in the
Atlantic Ocean some 357.6 nautical miles (412 sm, 665 km)
downrange from Cape Kennedy (latitude 30.24 degrees N
and longitude 74.016 degrees W) about nine minutes after
liftoff.

Second Stage

The second stage, with engines running 6 minutes
and 7 seconds will propel the vehicle to an altitude of
about 105.8 nautical miles (121.9 sm, 197 km) some 805
nautical miles (927.4 sm, 1490 km) downrange, building
up to 13,245 knots (15,258.3 mph) space fixed velocity.
The spent second stage will land in the Atlantic Ocean
about 19 minutes after lift-off some 2,190 nautical miles
(2,527 sm, 407 km) from the launch site, at latitude 31.79
degrees N and longitude 38.24 degrees W.

First Third-Stage Burn

The third stage, in its 152-second initial burn,
will place itself and the Apollo spacecraft in a circular
orbit 103 nautical miles (119 sm, 191.3 km) above the
Earth. Its inclination will be 32.5 degrees and orbital
period, 88.2 minutes. Apollo 8 will enter orbit at
about 47.08 degrees W longitude and 26.33 degrees
N latitude at a velocity of 25,592 feet-per-second
(17,433 statute mph or 15,132 knots).

- more -

Parking Orbit

While in the two revolutions in Earth parking orbit, the Saturn V third stage and spacecraft systems will be checked out in preparation for the second S-IVB burn.

Second Third-Stage Burn

Near the end of the second revolution, the J-2 engine of the third stage will be reignited for 5 minutes 12 seconds. This will inject the vehicle and spacecraft into a translunar trajectory. About 20 minutes later the CSM separates from the S-IVB/IU. Following separation, the S-IVB performs an attitude maneuver in preparation for dumping LOX residuals and a burn to depletion of the S-IVB auxiliary propulsion system (APS). Dumping of S-IVB LOX residuals and APS Burn on Apollo lunar missions may be done to alter the velocity and trajectory of the spent S-IVB-IU to place it in a "slingshot" trajectory passing behind the Moon's trailing edge into solar orbit.

Differences In Apollo 6 and Apollo 8 Launch Vehicles

The new helium prevalve cavity pressurization system will be flying on the S-IC for the first time. In this system, cavities in the liquid oxygen prevalves are filled with helium to create accumulators or "shock absorbers" to damp out oscillations. This system was installed to prevent excessive longitudinal oscillations experienced in the Apollo 6 flight.

The center engine of the S-IC stage will be cut off early (126 seconds after liftoff) in the boost phase. This is being done to keep acceleration forces from passing the four "g" level.

Software changes in the instrument unit will give a new cant capability to the outboard F-1 engines of the first stage. The engines will be in the normal position until the vehicle clears the launch tower. The engines will then cant outward two degrees to reduce the load on the spacecraft in the event one of the F-1 engines cut off prematurely.

- more -

The S-IC stage will also be carrying added instrumentation on this flight. The added instruments will monitor the actions of the first stage in connection with the accumulators installed to reduce longitudinal oscillations.

On the S-II stage, the propellant utilization system will have an open loop capability for improved reliability. Commands from the instrument unit will keep the propellants close to the planned levels.

The liquid hydrogen engine feed line for each J-2 engine has been redesigned, and the auxiliary spark igniter lines have been replaced.

Leaks developed in the Augmented Spark Igniter (ASI) lines of two J-2 engines on the second Saturn V launched. The new lines, without flex joints, were tested on the Apollo 7 flight and verified as satisfactory.

Film cameras will not be carried on the second stage of Apollo 8. On Apollo 6 two recoverable cameras recorded the first and second plane separations of the first and second stages.

A lightweight forward bulkhead in the liquid hydrogen tank is being used in this vehicle. The original bulkhead was damaged during production of the S-II stage about two years ago. The new bulkhead is the same type which will be used in S-II stages of all future Saturn V vehicles.

Added instrumentation will also be flown on the S-II stage. The added equipment will monitor the actions of the new ASI lines and further verify their suitability for future flights.

The thrust of the J-2 engines on this S-II stage will reach almost 230,000 pounds at one point. (Average engine thrust will be 228,290.7 pounds for a total stage thrust of 1,141,453.7 pounds.) The engines will ignite at a mixture ratio of 5.0 to 1 and then shift to 5.5 to 1 for the first portion of the burn. The high thrust occurs during the high mixture ratio burn phase. The mixture ratio then shifts at 440 seconds after liftoff back to 4.5 to 1 for the remainder of the powered flight. (Average engine thrust is 180,168.7 pounds for a stage total thrust of 900,843.7 pounds.)

- more -

The propellant utilization system for the S-IVB stage will also have the open loop capability.

The J-2 engine of the third stage is an uprated version with a top thrust of 230,000 pounds at the mixture ratio of 5.5 to 1. However, the engine will not reach 230,000 pounds thrust on this flight because the 5.5 to 1 mixture ratio is not planned. It is the first uprated J-2 engine to be flown.

A helium heater will be used as a repressurization system on the S-IVB for the first time on this flight. In this system, cold helium from the storage bottles are heated to provide pressurization for the fuel tank.

The capability to reignite the S-IVB stage engine after separation from the spacecraft has been added to this vehicle. The prime mission plan calls for the stage to be reignited in Earth parking orbit with the spacecraft attached. The added capability will permit stage reignition even if some unforeseen trouble requires separation of the spacecraft from the stage while in Earth parking orbit.

For example, loss of an F-1 engine during S-IC powered flight might require the use of too much S-IVB propellant in reaching Earth parking orbit. This would leave a shortage of propellant for the translunar injection burn. Also, a condition might be discovered during parking orbit which would pose a risk to the astronauts if the S-IVB was restarted while the spacecraft was attached. In any event of separation, the S-IVB will still be restarted, if possible, so that test data can be obtained from the stage even though it will be unmanned.

The redesigned ASI line will also be flown on this flight. It will be monitored again to verify satisfactory findings on the AS-205 flight. Instrumentation for checking out the new line has also been added, while research and development instrumentation no longer needed in some other parts of the vehicle has been removed.

Launch Vehicle Instrumentation and Communications

A total of 2,608 measurements will be taken in flight on the Saturn V launch vehicle.

- more -

This includes 893 in the first stage, 978 in the second stage, 359 in the third stage and 378 in the instrument unit.

The Saturn V will carry 20 telemetry systems, six on the first stage, six on the second stage, three on the third stage and five on the instrument unit. The vehicle will carry a radar tracking system on the first stage and a C-Band system and command system on the instrument unit. Each propulsive stage will carry a range safety system as on previous flights. Four motion picture cameras in recoverable capsules and two television cameras will also be on the first stage.

APOLLO 8 LAUNCH OPERATIONS

NASA's John F. Kennedy Space Center is responsible for preflight checkout, test and launch of the Apollo 8 space vehicle. A government-industry team of about 450 will conduct the final countdown from Firing Room 1 of the Launch Control Center (LCC).

The firing room team is backed up by more than 5,000 persons who are directly involved in launch operations at KSC--from the time the vehicle and spacecraft stages arrive at the center until the launch is completed.

Two major decisions made while flight hardware for Apollo 8 was at Kennedy Space Center had a pronounced affect on the scheduling of checkout and launch operations at KSC.

The first was the decision in April of this year to fly Apollo 8 as the first manned Saturn V flight. At that time, the launch vehicle had been erected on its mobile launcher in high bay number 1 of the Vehicle Assembly Building (VAB).

As a result of this decision, it was necessary to disassemble the stages so that the second stage (S-II) could be returned to Marshall Space Flight Center's Mississippi Test Facility (MTF). There it underwent modifications to prepare it for manned flight and to ensure that the problem which caused premature shutdown of two engines on the previous flight was corrected.

While the second stage was at MTF, modification kits were sent to KSC for installation on the first (S-IC) and third (S-IVB) stages to correct the launch vehicle longitudinal vibration anomaly and the failure of the third-stage engine to reignite in orbit. Both of these problems occurred on the second unmanned Saturn V flight, Apollo 6.

The second stage was returned to KSC in late June and erected on the booster in high bay number 1 in July. The third stage and instrument unit were erected the following month.

In August the decision was made not to fly a lunar module on Apollo 8. Lunar Module 3, which had been scheduled for the mission, had been at KSC since June, and was carried as part of the Apollo 8 spacecraft checkout schedule. As a result of this decision, LM-3 was dropped from the spacecraft schedule and a lunar module test article was inserted.

Assembly and checkout of Apollo 8 has been carried out while launch teams at KSC prepared and launched the Apollo 7 mission and began preparation for the Apollo 9 and Apollo 10 missions, scheduled for the first and second quarter of next year.

On Oct. 9, when the Apollo 8 space vehicle was moved from the Vehicle Assembly Building to Pad A, the launch team for Apollo 7 was conducting the final countdown to launch for the first manned Apollo mission. The Apollo 9 launch vehicle was already erected in the Vehicle Assembly Building and its command, service, and lunar modules were undergoing checkout in the Manned Spacecraft Operations Building (MSOB) in the industrial area. Lunar Module 4, the first flight hardware for Apollo 10 to arrive at KSC, was also undergoing checkout in the MSOB.

The Apollo 8 command and service modules arrived at KSC in mid-August and were taken to the Manned Spacecraft Operations Building for checkout and altitude chamber runs. The prime and backup crews each participated in altitude chamber tests during which spacecraft and crew systems were checked out at simulated altitudes in excess of 200,000 feet.

At the completion of testing in the altitude chamber, the CSM was mated to the spacecraft lunar module adapter (SLA) and moved to the VAB where it was mechanically mated to the launch vehicle. The move to the VAB and erection on the launch vehicle was completed Oct. 7. Following installation of certain ordnance items and the launch escape system, the space vehicle was moved to Pad A by the crawler transporter.

Integrated systems testing was conducted at the pad and the space vehicle was electrically mated in early November.

- more -

The first overall test of the space vehicle, called the Overall Test, Plugs In, verified the compatibility of the space vehicle systems, ground support equipment, and off-site support facilities by demonstrating the ability of the systems to proceed through a simulated countdown, launch, and flight. During the simulated flight portion of the test, the systems were required to respond to both emergency and normal flight conditions.

The space vehicle Flight Readiness Test was conducted in mid-November. This was the last overall test before the countdown demonstration. Both the prime and backup crews participate in portions of the FRT, which is a final overall test of the vehicle systems and associated ground support equipment when all systems are as near as possible to a launch configuration.

After hypergolic fuels were loaded aboard the space vehicle and RP-1, the launch vehicle first stage fuel was brought aboard, and the final major test of the space vehicle began. This was the countdown demonstration test (CDDT), a dress rehearsal for the final countdown to launch. The CDDT for Apollo 8 was divided into a "wet" and a "dry" portion. During the first, or "wet" portion, the entire countdown, including propellant loading, was carried out down to T-8.9 seconds. The astronaut crews did not participate in the wet CDDT. At the completion of the wet CDDT, the cryogenic propellants (liquid oxygen and liquid hydrogen) were off-loaded, and the final portion of the countdown was re-run, this time simulating the fueling and with the prime astronaut crew participating as they will on launch day.

Because of the complexity involved in the checkout of the 363-foot-tall Apollo/Saturn V configuration, the launch teams make use of extensive automation in their checkout. Automation is one of the major differences in checkout used on Apollo compared to the procedures used in the Mercury and Gemini programs.

RCA 110A computers, data display equipment and digital data techniques are used throughout the automatic checkout from the time the launch vehicle is erected in the VAB through liftoff. A similar, but separate computer operation called ACE (Acceptance Checkout Equipment) is used to verify the flight readiness of the spacecraft. Spacecraft checkout is controlled from separate firing rooms located in the Manned Spacecraft Operations Building.

Countdown

Hrs: Min: Secs:

T - 28:	00:	00	Start launch vehicle and spacecraft countdown activities
- 24:	30:	00	S-II power-up
- 24:	00:	00	S-IVB power-up
- 09:	00:	00	Six-hour built-in hold
- 09:	00:	00	End of built-in hold close Command Module and boost protective cover hatch
- 08:	59:	00	Clear pad for launch vehicle cryo loading
- 07:	28:	00	Start S-IVB LOX loading
- 07:	04:	00	S-IVB LOX loading complete; start S-II LOX loading
- 06:	27:	00	S-II LOX loading complete; start S-IC LOX loading
- 04:	57:	00	S-IC LOX loading complete
- 04:	54:	00	Start S-II liquid hydrogen loading
- 04:	11:	00	S-II liquid hydrogen loading complete; start S-IVB liquid hydrogen loading
- 03:	30:	00	Flight crew departs Manned Spacecraft Operations building
- 03:	28:	00	S-IVB liquid hydrogen loading complete
- 03:	13:	00	Closeout crew on station; start ingress preps
- 02:	40:	00	Start flight crew ingress
- 02:	10:	00	Flight crew ingress complete

- more -

Hrs: Min: Secs:

T - 01:	40:	00		Close spacecraft hatch
- 01:	00:	00		Start RP-1 level adjust
- 00:	42:	00		Arm Launch Escape System pyro buses
- 00:	35:	00		RP-1 level adjust complete
- 00:	15:	00		Spacecraft on internal power
- 00:	05:	30		Arm Safe and Arm devices
- 00:	03:	07		Terminal Count Sequence (TCS) start
- 00:	00:	17.2		Guidance reference release command
- 00:	00:	08.9		S-IC ignition command
- 00:	00:	00		Lift-off

NOTE: The foregoing is the Apollo 8 countdown that was prepared as the press kit was ready for printing. Some
changes may be made as a result of the Countdown
Demonstration Test (CDDT).

KSC Launch Complex 39

Launch Complex 39 facilities at the Kennedy Space Center were planned and built specifically for the Saturn V program, the space vehicle that will be used to carry astronauts to the Moon.

Complex 39 introduced the mobile concept of launch operations, a departure from the fixed launch pad techniques used previously at Cape Kennedy and other launch sites. Since the early 1950's when the first ballistic missiles were launched, the fixed launch concept had been used on NASA missions. This method called for assembly, checkout and launch of a rocket at one site--the launch pad. In addition to tying up the pad, this method also often left the flight equipment exposed to the outside influences of the weather for extended periods.

Using the mobile concept, the space vehicle is thoroughly checked in an enclosed building before it is moved to the launch pad for final preparations. This affords greater protection, a more systematic checkout process using computer techniques, and a high launch rate for the future, since the pad time is minimal.

Saturn V stages are shipped to the Kennedy Space Center by ocean-going vessels and specially designed aircraft, such as the Guppy. Apollo spacecraft modules are transported by air. The spacecraft components are first taken to the Manned Spacecraft Operations Building for preliminary checkout. The Saturn V stages are brought immediately to the Vehicle Assembly Building after arrival at the nearby turning basin.

Apollo 8 is the third Saturn V to be launched from Pad A, Complex 39. The historic first launch of the Saturn V, designated Apollo 4, took place Nov. 9, 1967 after a perfect countdown and on-time liftoff at 7 a.m. EST. The second Saturn V mission--Apollo 6--was conducted last April 4.

The major components of Complex 39 include: (1) the Vehicle Assembly Building (VAB) where the Apollo 8 was assembled and prepared; (2) the Launch Control Center, where the launch team conducts the preliminary checkout and countdown; (3) the mobile launcher, upon which the Apollo 8 was erected for checkout and from where it will be launched; (4) the mobile service structure, which provides external access to the space vehicle at the pad; (5) the transporter, which carries the space vehicle and mobile launcher, as well as the mobile service structure to the pad; (6) the crawlerway over which the space vehicle travels from the VAB to the launch pad; and (7) the launch pad itself.

The Vehicle Assembly Building

The Vehicle Assembly Building is the heart of Launch Complex 39. Covering eight acres, it is where the 363-foot-tall space vehicle is assembled and tested.

The VAB contains 129,482,000 cubic feet of space. It is 716 feet long, and 518 feet wide and it covers 343,500 square feet of floor space.

The foundation of the VAB rests on 4,225 steel pilings, each 16 inches in diameter, driven from 150 to 170 feet to bedrock. If placed end to end, these piles would extend a distance of 123 miles. The skeletal structure of the building contains approximately 60,000 tons of structural steel. The exterior is covered by more than a million square feet of insulated aluminum siding.

The building is divided into a high bay area 525 feet high and a low bay area 210 feet high, with both areas serviced by a transfer aisle for movement of vehicle stages.

The low bay work area, approximately 442 feet wide and 274 feet long, contains eight stage-preparation and checkout cells. These cells are equipped with systems to simulate stage interface and operation with other stages and the instrument unit of the Saturn V launch vehicle.

After the Apollo 8 launch vehicle upper stages arrived at the Kennedy Space Center, they were moved to the low bay of the VAB. Here, the second and third stages underwent acceptance and checkout testing prior to mating with the S-IC first stage atop mobile launcher No. 1 in the high bay area.

The high bay provides the facilities for assembly and checkout of both the launch vehicle and spacecraft. It contains four separate bays for vertical assembly and checkout. At present, three bays are equipped, and the fourth will be reserved for possible changes in vehicle configuration.

Work platforms -- some as high as three-story buildings --
in the high bays provide access by surrounding the launch vehicle
at varying levels. Each high bay has five platforms. Each
platform consists of two bi-parting sections that move in from
opposite sides and mate, providing a 360-degree access to the
section of the space vehicle being checked.

A 10,000-ton-capacity air conditioning system, sufficient
to cool about 3,000 homes, helps to control the environment
within the entire office, laboratory, and workshop complex
located inside the low bay area of the VAB. Air conditioning
is also fed to individual platform levels located around the
vehicle.

There are 141 lifting devices in the VAB, ranging from
one-ton hoists to two 250-ton high-lift bridge cranes.

The mobile launchers, carried by transporter vehicles,
move in and out of the VAB through four doors in the high
bay area, one in each of the bays. Each door is shaped like
an inverted T. They are 152 feet wide and 114 feet high at
the base, narrowing to 76 feet in width. Total door height is
456 feet.

The lower section of each door is of the aircraft
hangar type that slides horizontally on tracks. Above this
are seven telescoping vertical lift panels stacked one above
the other, each 50 feet high and driven by an individual motor.
Each panel slides over the next to create an opening large
enough to permit passage of the Mobile Launcher.

The Launch Control Center

Adjacent to the VAB is the Launch Control Center (LCC).
This four-story structure is a radical departure from the
dome-shaped blockhouses at other launch sites.

The electronic "brain" of Launch Complex 39, the LCC
was used for checkout and test operations while Apollo 8 was
being assembled inside the VAB. The LCC contains display,
monitoring, and control equipment used for both checkout and
launch operations.

The building has telemeter checkout stations on its
second floor, and four firing rooms, one for each high bay
of the VAB, on its third floor. Three firing rooms will contain
identical sets of control and monitoring equipment, so that
launch of a vehicle and checkout of others may take place
simultaneously. A ground computer facility is associated with
each firing room.

The high speed computer data link is provided between the LCC and the mobile launcher for checkout of the launch vehicle. This link can be connected to the mobile launcher at either the VAB or at the pad.

The three equipped firing rooms have some 450 consoles which contain controls and displays required for the checkout process. The digital data links connecting with the high bay areas of the VAB and the launch pads carry vast amounts of data required during checkout and launch.

There are 15 display systems in each LCC firing room, with each system capable of providing digital information instantaneously.

Sixty television cameras are positioned around the Apollo/Saturn V transmitting pictures on 10 modulated channels. The LCC firing room also contains 112 operational intercommunication channels used by the crews in the checkout and launch countdown.

Mobile Launcher

The mobile launcher is a transportable launch base and umbilical tower for the space vehicle. Three launchers are used at Complex 39.

The launcher base is a two-story steel structure, 25 feet high, 160 feet long, and 135 feet wide. It is positioned on six steel pedestals 22 feet high when in the VAB or at the launch pad. At the launch pad, in addition to the six steel pedestals, four extendable colums also are used to stiffen the mobile launcher against rebound loads, if the engine cuts off.

The umbilical tower, extending 398 feet above the launch platform, is mounted on one end of the launcher base. A hammerhead crane at the top has a hook height of 376 feet above the deck with a traverse radius of 85 feet from the center of the tower.

The 12-million-pound mobile launcher stands 445 feet high when resting on its pedestals. The base, covering about half an acre, is a compartmented structure built of 25-foot steel girders.

The launch vehicle sits over a 45-foot-square opening which allows an outlet for engine exhausts into a trench containing a flame deflector. This opening is lined with a replaceable steel blast shield, independent of the structure, and will be cooled by a water curtain initiated two seconds after liftoff.

There are nine hydraulically-operated service arms on the umbilical tower. These swing arms support lines for the vehicle umbilical systems and provide access for personnel to the stages as well as the astronaut crew to the spacecraft.

On Apollo 8 two of the service arms (including the Apollo spacecraft access arm) are retracted early in the count. A third is released at T-30 seconds, and a fourth at about T-15 seconds. The remaining five arms are set to swing back at vehicle first motion after T-0.

The swing arms are equipped with a backup retraction system in case the primary mode fails.

The Apollo access arm (swing arm No. 9), located at the 320-foot level above the launcher base, provides access to the spacecraft cabin for the closeout team and astronaut crews. Astronauts Borman, Lovell and Anders will board the spacecraft starting at about T-2 hours, 40 minutes in the count. The access arm will be moved to a parked position, 12 degrees from the spacecraft, at about T-42 minutes.

This is a distance of about three feet, which permits a rapid reconnection of the arm to the spacecraft in the event of an emergency condition. The arm is fully retracted at the T-5 minute mark in the count.

The Apollo 8 vehicle is secured to the mobile launcher by four combination support and hold-down arms mounted on the launcher deck. The hold-down arms are cast in one piece, about 6 by 9 feet at the base and 10 feet tall, weighing more than 20 tons. Damper struts secure the vehicle near its top.

After the engines ignite, the arms hold Apollo 8 for about six seconds until the engines build up to 95 per cent thrust and other monitored systems indicate they are functioning properly. The arms release on receipt of a launch commit signal at the zero mark in the count.

The Transporter

The six-million-pound transporters, the largest tracked vehicles known, move mobile launchers into the VAB and mobile launchers with assembled Apollo space vehicles to the launch pad. They also are used to transfer the mobile service structure to and from the launch pads. Two transporters are in use at Complex 39

The Transporter is 131 feet long and 114 feet wide. The vehicle moves on four double-tracked crawlers, each 10 feet high and 40 feet long. Each shoe on the crawler tracks seven feet six inches in length and weighs about a ton.

Sixteen traction motors powered by four 1,000-kilowatt generators, which in turn are driven by two 2,750-horsepower diesel engines, provide the motive power for the transporter. Two 750-kw generators, driven by two 1,065-horsepower diesel engines, power the jacking, steering, lighting, ventilating and electronic systems.

Maximum speed of the transporter is about one-mile-per-hour loaded and about two-miles-per-hour unloaded. A 3½ mile trip to the pad with a mobile launcher, made at less than maximum speed, takes approximately seven hours.

The transporter has a leveling system designed to keep the top of the space vehicle vertical within plus-or-minus 10 minutes of arc -- about the dimensions of a basketball.

This system also provides leveling operations required to negotiate the five per cent ramp which leads to the launch pad, and keeps the load level when it is raised and lowered on pedestals both at the pad and within the VAB.

The overall height of the transporter is 20 feet from ground level to the top deck on which the mobile launcher is mated for transportation. The deck is flat and about the size of a baseball diamond (90 by 90 feet).

Two operator control cabs, one at each end of the chassis located diagonally opposite each other, provide totally enclosed stations from which all operating and control functions are coordinated.

The transporter moves on a roadway 131 feet wide, divided by a median strip. This is almost as broad as an eight-lane turnpike and is designed to accommodate a combined weight of about 18 million pounds.

The roadway is built in three layers with an average depth of seven feet. The roadway base layer is two-and-one-half feet of hydraulic fill compacted to 95 per cent density. The next layer consists of three feet of crushed rock packed to maximum density, followed by a layer of one foot of selected hydraulic fill. The bed is topped and sealed with an asphalt prime coat.

On top of the three layers is a cover of river rock, eight inches deep on the curves and six inches deep on the straightway. This layer reduces the friction during steering and helps distribute the load on the transporter bearings.

Mobile Service Structure

A 402-foot-tall, 9.8-million-pound tower is used to service the Apollo-launch vehicle and spacecraft at the pad. The 40-story steel-trussed tower, called a mobile service structure, provides 360-degree platform access to the Saturn vehicle and the Apollo spacecraft.

The service structure has five platforms -- two self-propelled and three fixed, but movable. Two elevators carry personnel and equipment between work platforms. The platforms can open and close around the 363-foot space vehicle.

After depositing the mobile launcher with its space vehicle on the pad, the transporter returns to a parking area about 7,000 feet from the pad. There it picks up the mobile service structure and moves it to the launch pad. At the pad, the huge tower is lowered and secured to four mount mechanisms.

The top three work platforms are located in fixed positions which serve the Apollo spacecraft. The two lower movable platforms serve the Saturn V.

The mobile service structure ramains in position until about T-11 hours when it is removed from its mounts and returned to the parking area.

Water Deluge System

A water deluge system will provide a million gallons of industrial water for cooling and fire prevention during launch of Apollo 8. Once the service arms are retracted at liftoff, a spray system will come on to cool these arms from the heat of the five Saturn F-1 engines during liftoff.

On the deck of the mobile launcher are 29 water nozzles. This deck deluge will start immediately after liftoff and will pour across the face of the launcher for 30 seconds at the rate of 50,000 gallons-per-minute. After 30 seconds, the flow will be reduced to 20,000 gallons-per-minute.

-more-

Positioned on both sides of the flame trench are a series of nozzles which will begin pouring water at 8,000 gallons-per-minute, 10 seconds before liftoff. This water will be directed over the flame deflector.

Other flush mounted nozzles, positioned around the pad, will wash away any fluid spill as a protection against fire hazards.

Water spray systems also are available along the egress route that the astronauts and closeout crews would follow in case an emergency evacuation was required.

Flame Trench and Deflector

The flame trench is 58 feet wide and approximately six feet above mean sea level at the base. The height of the trench and deflector is approximately 42 feet.

The flame deflector weighs about 1.3 million pounds and is stored outside the flame trench on rails. When it is moved beneath the launcher, it is raised hydraulically into position. The deflector is covered with a four-and-one-half-inch thickness of refractory concrete consisting of a volcanic ash aggregate and a calcuim aluminate binder. The heat and blast of the engines are expected to wear about three-quarters of an inch from this refractory surface during the Apollo 8 launch.

Pad Areas

Both Pad A and Pad B of Launch Complex 39 are roughly octagonal in shape and cover about one fourth of a square mile of terrain.

The center of the pad is a hardstand constructed of heavily reinforced concrete. In addition to supporting the weight of the mobile launcher and the Saturn V vehicle, it also must support the 9.8-million-pound mobile service structure and 6-million-pound transporter, all at the same time. The top of the pad stands some 48 feet above sea level.

Saturn V propellants -- liquid oxygen, liquid hydrogen, and RP-1 -- are stored near the pad perimeter.

Stainless steel, vacuum-jacketed pipes carry the liquid oxygen (LOX) and liquid hydrogen from the storage tanks to the pad, up the mobile launcher, and finally into the launch vehicle propellant tanks.

LOX is supplied from a 900,000-gallon storage tank. A centrifugal pump with a discharge pressure of 320 pounds-per-square-inch pumps LOX to the vehicle at flow rates as high as 10,000-gallons-per-minute.

Liquid hydrogen, used in the second and third stages, is stored in an 850,000-gallon tank, and is sent through 1,500 feet of 10-inch, vacuum-jacketed invar pipe. A vaporizing heat exchanger pressurizes the storage tank to 60 psi for a 10,000-gallons-per-minute flow rate.

The RP-1 fuel, a high grade of kerosene is stored in three tanks--each with a capacity of 86,000 gallons. It is pumped at a rate of 2,000 gallons-per-minute at 175 psig.

The Complex 39 pneumatic system includes a converter-compressor facility, a pad high-pressure gas storage battery, a high-pressure storage battery in the VAB, low and high-pressure, cross-country supply lines, high-pressure hydrogen storage and conversion equipment, and pad distribution piping to pneumatic control panels. The various purging systems require 187,000 pounds of liquid nitrogen and 21,000 gallons of helium.

MISSION CONTROL CENTER

The Mission Control Center at the Manned Spacecraft Center, Houston, is the focal point for all Apollo flight control activities. The Center will receive tracking and telemetry data from the Manned Space Flight Network. These data will be processed through the Mission Control Center Real-Time Computer Complex and used to drive displays for the flight controllers and engineers in the Mission Operations Control Room and staff support rooms.

The Manned Space Flight Network tracking and data acquisition stations link the flight controllers at the Center to the spacecraft.

For Apollo 8, all stations will be remote sites without flight control teams. All uplink commands and voice communications will originate from Houston, and telemetry data will be sent back to Houston at high speed (2,400 bits per second), on two separate data lines. They can be either real time or playback information.

Signal flow for voice circuits between Houston and the remote sites is via commercial carrier, usually satellite, wherever possible using leased lines which are part of the NASA Communications Network.

Commands are sent from Houston to NASA's Goddard Space Flight Center, Greenbelt, Md., lines which link computers at the two points. The Goddard computers provide automatic switching facilities and speed buffering for the command data. Data are transferred from Goddard to remote sites on high speed (2,400 bits per second) lines. Command loads also can be sent by teletype from Houston to the remote sites at 100 words per minute. Again, Goddard computers provide storage and switching functions.

Telemetry data at the remote site are received by the RF receivers, processed by the Pulse Code Modulation ground stations, and transferred to the 642B remote-site telemetry computer for storage. Depending on the format selected by the telemetry controller at Houston, the 642B will output the desired format through a 2010 data transmission unit which provides parallel to serial conversion, and drives a 2,400 bit-per-second modem.

The data modem converts the digital serial data to phase-shifted keyed tones which are fed to the high speed data lines of the Communications Network.

Telemetry summary messages can also be output by the 642B computer, but these messages are sent to Houston on 100-word-per-minute teletype lines rather than on the high-speed lines.

Tracking data are output from the sites in a low speed (100 words) teletype format and a 240-bit block high speed (2,400 bits) format. Data rates are 1 sample-6 seconds for teletype and 10 samples (frames) per second for high speed data.

All high-speed data, whether tracking or telemetry, which originate at a remote site are sent to Goddard on high-speed lines. Goddard reformats the data when necessary and sends them to Houston in 600-bit blocks at a 40,800 bits-per-second rate. Of the 600-bit block, 480 bits are reserved for data, the other 120 bits for address, sync, intercomputer instructions, and polynominal error encoding.

All wideband 40,800 bits-per-second data originating at Houston are converted to high speed (2,400 bits-per-second) data at Goddard before being transferred to the designated remote site.

MANNED SPACE FLIGHT NETWORK

.

The Manned Space Flight Network (MSFN) will have 14 ground stations, four instrumented ships, and six instrumented aircraft ready for participation in Apollo 8.

The MSFN is designed to keep in close contact with the spacecraft and astronauts at all times, except for the approximate 45 minutes Apollo will be behind the Moon. The network is designed to provide reliable, continuous, and instantaneous communications with the astronauts, launch vehicle, and spacecraft from liftoff to splashdown.

As the spacecraft lifts off from Kennedy Space Center, the tracking stations will be watching it. As the Saturn ascends, voice and data will be instantaneously transmitted to Mission Control Center (MCC) in Houston. Data will be run through computers at MCC for visual display for flight controllers.

Depending on the launch azimuth, a string of 30-foot diameter antennas around the Earth will keep tabs on Apollo 8 and transmit information back to Houston. First, the station at Merritt Island, then it will be Grand Bahama Island, Bermuda, the Vanguard tracking ship, and Canary Island. Later, Carnarvon, Australia, will pick up Apollo 8, followed by Hawaii, the Redstone tracking ship, Guaymas, Mexico, and Corpus Christi, Texas.

For injection into translunar orbit, MCC sends a signal through one of the land stations or one of the three Apollo ships in the Pacific. As the spacecraft heads for the Moon, the engine burn is monitored by the ships and an Apollo/Range Instrumentation Aircraft (A/RIA). The A/RIA provides a relay for the astronauts' voice and data communication with Houston.

As the spacecraft moves away from Earth, first the smaller 30-foot diameter antennas communicate with the spacecraft, then at a spacecraft altitude of 10,000 miles they hand over the tracking function to the larger and more powerful 85-foot antennas. These 85-foot antennas are near Madrid, Spain; Goldstone, Calif.; and Canberra, Australia.

The 85-foot antennas are at 120-degree intervals around Earth so at least one antenna has the Moon in view at all times. As the Earth revolves from west to east, one station hands over control to the next station as it moves into view of the spacecraft. In this way, a continuous data and communication flow is maintained.

-more-

Data is constantly relayed back through the huge antennas and transmitted via the NASA Communications Network--a half million miles of land and underseas cables and radio circuits, including those through communications satellites--to MCC. This data is fed into computers for visual display in Mission Control. For example, a display would show on a large map, the exact position of the spacecraft. Or returning data could indicate a drop in power or some other difficulty which would result in a red light going on to alert a Flight Controller to make a decision and take action.

Returning data flowing through the Earth stations give the necessary information for commanding mid-course maneuvers to keep the Apollo in a proper trajectory for orbiting the Moon. On reaching the vicinity of the Moon the data indicate the amount of burn necessary for the service module engine to place the spacecraft in lunar orbit. And so it goes, continuous tracking and acquisition of data between Earth and Apollo are used to fire the spacecraft's engine to return home and place it on the precise trajectory for reentering the Earth's atmosphere.

As the spacecraft comes toward Earth at about 25,000 miles per hour, it must reenter at the proper angle.

Calculations based on data coming in at the various tracking stations and ships are fed into the computers at MCC where flight controllers make decisions that will provide the returning spacecraft with the necessary information to make accurate reentry. Appropriate MSFN stations, including tracking ships and aircraft repositioned in the Pacific for this event, are on hand to provide support during reentry. An A/RIA aircraft will relay astronaut voice communications to MCC and antennas on reentry ships will follow the spacecraft.

During the journey to the Moon and back, television will be received from the spacecraft at the various 85-foot antennas around the world: Spain, Goldstone, and Australia. Scan converters at Madrid and Goldstone permit immediate transmission via NASCOM to Mission Control where it will be released to TV networks.

NASA Communications Network - Goddard

This network consists of several systems of diversely routed communications channels leased on communications satellites, common carrier systems and high frequency radio facilities where necessary to provide the access links.

The system consists of both narrow and wide-band channels, and some TV channels. Included are a variety of telegraph, voice and data systems (digital and analog) with a wide range of digital data rates. Wide-band systems do not extend overseas. Alternate routes or redundancy are provided for added reliability in critical mission operations.

A primary switching center and intermediate switching and control points are established to provide centralized facility and technical control, and switching operations under direct NASA control. The primary switching center is at Goddard, and intermediate switching centers are located at Canberra, Australia; Madrid, Spain; London, England; Honolulu, Hawaii; Guam; and Cape Kennedy, Florida.

For Apollo 8, Cape Kennedy is connected directly to the Mission Control Center, Houston, by the communication network's Apollo Launch Data System, a combination of data gathering and transmission systems designed to handle launch data exclusively.

After launch, all network and tracking data are directed to the Mission Control Center, Houston, through Goddard. A high-speed data line (2,400 bits-per-second) connects Cape Kennedy to Goddard, where the transmission rate is increased to 40,800 bits-per-second from there to Houston. Upon orbital insertion, tracking responsibility is transferred between the various stations as the spacecraft circles the Earth.

Two Intelsat communications satellites will be used for Apollo 8. The Atlantic satellite will service the Ascension Island Unified S-Band (USB) station, the Atlantic Ocean ship and the Canary Island site.

The second Apollo Intelsat communications satellite, over the mid-Pacific, will service the Carnarvon, Australia USB site and the Pacific Ocean ships. All these stations will be able to transmit simultaneously through the satellite to Houston via Brewster Flat, Washington, and the Goddard Space Flight Center.

Network Computers

At fraction-of-a-second intervals, the network's digital data processing systems, with NASA's Manned Spacecraft Center as the focal point, "talk" to each other or to the spacecraft in real time. High-speed computers at the remote site (tracking ships included) issue commands or "up" data on such matters as control of cabin pressure, orbital guidance commands, or "go-no-go" indications to perform certain functions.

In the case of information originating from Houston, the computers refer to their pre-programmed information for validity before transmitting the required data to the capsule.

Such "up" information is communicated by ultra-high-frequency radio at about 1,200 bits-per-second. Communication between remote ground sites, via high-speed communications links, occurs about the same rate. Houston reads information from these ground sites at 2,400 bits-per-second, as well as from remote sites at 100 words-per-minute.

The computer systems perform many other functions, including:

> Assuring the quality of the transmission lines by continually exercising data paths.
>
> Verifying accuracy of the messages by repetitive operations.
>
> Constantly updating the flight status.

For "down" data, sensors built into the spacecraft continually sample cabin temperature, pressure, physical information on the astronauts such as heartbeat and respiration, among other items. These data are transmitted to the ground stations at 51.2 kilobits (12,800 binary digits) per second.

At MCC the computers:

> Detect and select changes or deviations, compare with their stored programs, and indicate the problem areas or pertinent data to the flight controllers.
>
> Provide displays to mission personnel.
>
> Assemble output data in proper formats.
>
> Log data on magnetic tape for replay.
>
> Provide storage for "on-call" display for the flight controllers.
>
> Keep time.

-more-

Fourteen land stations are outfitted with computer systems to relay telemetry and command information between Houston and Apollo spacecraft: Canberra and Carnarvon, Australia; Guam; Kauai, Hawaii; Goldstone, California; Corpus Christi, Texas; Cape Kennedy, Florida; Grand Bahama Island; Bermuda; Madrid; Grand Canary Island; Antigua; Ascension Island; and Guaymas, Mexico.

Network Configuration for Apollo 8

Unified S-Band (USB) Sites:

NASA 30-Foot Antenna Sites	NASA 85-Foot Antenna Sites
Antigua (ANG)	Canberra (CNB), Australia (Prime)
Ascension Island (ACN)	
Bermuda (BDA)	Goldstone (GDS), California (Prime)
Canary Island (CYI)	
Carnarvon (CRO), Australia	Madrid (MAD), Spain (Prime)
Grand Bahama Island (GBM)	*Canberra (DSS-42 Apollo Wing) (Backup)
Guam (GWM)	
Guaymas (GYM), Mexico	*Goldstone (DSS-11 Apollo Wing) (Backup)
Hawaii (HAW)	
Merritt Island (MIL), Florida	*Madrid (DSS-61 Apollo Wing) (Backup)
Corpus Christi (TEX), Texas	

Tananarive (TAN), Malagasy Republic (STADAN station in support role only.)

---

*Wings have been added to JPL Deep Space Network site operations buildings. These wings contain additional Unified S-Band equipment as backup to the Prime sites.

MANNED SPACE FLIGHT NETWORK (APOLLO-8)

NASCOM-APOLLO 8

APOLLO 8 RECOVERY

The Apollo Ships

The mission will be supported by four Apollo instrumentation ships operating as integral stations of the Manned Space Flight Network (MSFN) to provide coverage in areas beyond the range of land stations.

The ships, Vanguard, Redstone, Mercury, and Huntsville will perform tracking, telemetry, and communication functions for the launch phase, Earth orbit insertion, translunar injection (TLI), and reentry at the end of the mission.

Vanguard will be stationed about 1,000 miles southeast of Bermuda (25°N, 49°W) to bridge the Bermuda-Antigua gap during Earth orbit insertion. Vanguard also functions as part of the Atlantic recovery fleet in the event of a launch phase contingency. Redstone, in the western Pacific, north of Bougainville (2.5°N, 155.5°E); Mercury, 1500 miles further east (7.5°N, 181.5°E); and Huntsville, near Wake Island (21.0°N, 169.0°E), provide a triangle of mobile stations between the MSFN stations at Carnarvon and Hawaii for coverage of the burn interval for translunar injection. In the event the launch date slips from December 21, the ships will all move generally southwestward to cover the changing flight window patterns.

Mercury and Huntsville will be repositioned along the reentry corridor for tracking, telemetry, and communications functions during reentry and landing.

The Apollo ships were developed jointly by NASA and the Department of Defense. The DOD operates the ships in support of Apollo and other NASA and DOD missions on a non-interference basis with Apollo requirements.

The overall management of the Apollo ships is the responsibility of the Commander, Air Force Western Test Range (AFWTR). The Military Sea Transport Service provides the maritime crews and the Federal Electric Corporation of International Telephone and Telegraph, under contract to AFWTR, provides the technical instrumentation crews.

- more -

The technical crews operate in accordance with joint NASA/DOD standards and specifications which are compatible with MSFN operational procedures.

Apollo/Range Instrumentation Aircraft (A/RIA)

The Apollo/Range Instrumentation Aircraft (A/RIA) will support the mission by filling gaps in both land and ship station coverage where important and significant coverage requirements exist.

During Apollo 8, the A/RIA will be used primarily to fill coverage gaps of the land and ship stations in the Pacific during the translunar injection interval (TLI). Prior to and during the TLI burn, the A/RIA record telemetry data from Apollo and provide a real-time voice communication between the astronauts and the flight director at Houston.

Six aircraft will participate in this mission flying from Pacific air bases to positions under the orbital track of the spacecraft and booster.

The A/RIA will fly, initially, out of Hawaii, Guam, and the Philippines, as well as three bases in Australia: Townsville, Darwin, and Perth. The aircraft, like the tracking ships, will also be redeployed in a southwest direction in the event of launch day slips.

The total A/RIA fleet for Apollo missions consist of eight EC-135-A (Boeing 707) jet aircraft equipped specifically to meet mission needs. Seven-foot parabolic antennas have been installed in the nose section of the aircraft giving them a large, bulbous look.

They are under the overall supervision of the Office of Tracking and Data Acquisition with direct supervision the responsibility of Goddard. The aircraft, as well as flight and instrumentation crews, are provided by the Air Force and they are equipped through joint Air Force-NASA contract action.

NETWORK CONFIGURATION FOR APOLLO 8 MISSION

Facilities	Tracking: C-band (High Speed)	C-band (Low Speed)	ODOP	Optical	USB: USB	Voice (A/G)	Command	Telemetry	TLM: VHF Links	FM Remoting	Mag Tape Recording	Decoms	Displays	CMD: CMD Destruct	Data Processing: 642B TLM	642B CMD	121s	Comm: High Speed Data	Wideband Data	TTY	Voice (SCAMA)	VHF A/G Voice	Video (TV)	Other: SPAN
CIF												X	X					X	X		X		X	
TEL 4									X	X	X													
CNV	X		X	X										X									X	
PAT*	X	X																						
MLA	X	X																						
MIL					X	X	X	X	X	X	X	X	X		X	X	X	X		X	X	X	X	
GBI	X*	X*				X				X				X	X	X	X	X		X	X	X	X	
GBM					X	X	X	X	X	X	X	X			X	X	X	X		X	X	X	X	
GTK	X	X												X										
ANG					X	X	X	X	X	X	X	X			X	X	X	X		X	X	X	X	
ANT	X	X								X	X													
BDA	X	X			X	X	X	X	X	X	X	X	X	X	X	X	X	X		X	X	X	X	
ACN					X	X	X	X	X	X	X	X			X	X	X	X		X	X	X	X	
ASC		X																						
MAD					X	X	X	X		X	X	X			X	X	X	X		X	X		X	
MADX					X	X	X	X									X			X	X			
CYI		X			X	X	X	X	X	X	X	X			X	X	X	X		X	X	X	X	X
PRE		X																						
TAN		X							X		X									X	X	X		
CRO	X	X			X	X	X	X	X	X	X	X			X	X	X	X		X	X	X	X	X
HSK					X	X	X	X		X	X	X			X	X	X	X		X	X		X	
HSKX					X	X	X	X									X			X	X			
GWM					X	X	X	X	X	X	X	X			X	X	X	X		X	X	X	X	
HAW		X			X	X	X	X	X	X	X	X	X		X	X	X	X		X	X	X	X	
CAL	X	X																		X	X	X		
WHS	X	X																		X	X			
GDS					X	X	X	X		X	X	X			X	X	X	X		X	X		X	
GDSX					X	X	X	X									X			X	X			
GYM					X	X	X	X	X	X	X	X			X	X	X	X		X	X	X	X	
TEX					X	X	X	X	X	X	X	X			X	X	X			X	X	X	X	
HTV		X			X	X		X			X	X								X	X	X		
RED	X	X			X	X	X	X	X	X	X	X			X	X		X	X	X	X	X	X	X
VAN	X	X			X	X	X	X	X	X	X				X	X		X		X	X	X	X	
MER	X	X			X	X	X	X	X	X	X				X	X		X		X	X	X	X	
ARIA (6)					X	X		X	X		X										X	X		

*Subject to availability.

APOLLO 8 CREW

Crew Training

The crewmen of Apollo 8 have spent more than seven hours of formal crew training for each hour of the lunar-orbit mission's six-day duration. Almost 1,100 hours of training were in the Apollo 8 crew training syllabus over and above the normal preparations for the mission-- technical briefings and reviews, pilot meetings and study.

The Apollo 8 crewmen also participated in spacecraft manufacturing checkouts at the North American Rockwell plant in Downey, Calif., and in prelaunch testing at NASA Kennedy Space Center. Taking part in factory and launch area testing has provided the crew with valuable operational knowledge of the complex vehicle.

Highlights of specialized Apollo 8 crew training topics are:

* Detailed series of briefings on spacecraft systems, operation and modifications.

* Saturn launch vehicle briefings on countdown, range safety, flight dynamics, failure modes and abort conditions. The launch vehicle briefings were updated periodically.

* Apollo Guidance and Navigation system briefings at the Massachusetts Institute of Technology Instrumentation Laboratory.

* Briefings and continuous training on mission photographic objectives and use of camera equipment.

* Extensive pilot participation in reviews of all flight procedures for normal as well as emergency situations.

* Stowage reviews and practice in training sessions in the spacecraft, mockups, and Command Module simulators allowed the crewmen to evaluate spacecraft stowage of crew-associated equipment.

- more -

* More than 200 hours of training per man in Command Module simulators at MSC and KSC, including closed-loop simulations with flight controllers in the Mission Control Center. Other Apollo simulators at various locations were used extensively for specialized crew training.

* Entry corridor deceleration profiles at lunar-return conditions in the MSC Flight Acceleration Facility manned centrifuge.

* Water egress training conducted in indoor tanks as well as in the Gulf of Mexico, included uprighting from the Stable II position (apex down) to the Stable I position (apex up), egress onto rafts and helicopter pickup.

* Launch pad egress training from mockups and from the actual spacecraft on the launch pad for possible emergencies such as fire, contaminants and power failures.

* The training covered use of Apollo spacecraft fire suppression equipment in the cockpit.

* Planetarium reviews at Morehead Planetarium, Chapel Hill, N. C., and at Griffith Planetarium, Los Angeles, Calif., of the celestial sphere with special emphasis on the 37 navigational stars used by the Command Module Computer.

Apollo 8 Spacesuits

Apollo 8 crewmen, until one hour after translunar injection, will wear the intravehicular pressure garment assembly--a multi-layer spacesuit consisting of a helmet, torso and gloves which can be pressurized independently of the spacecraft.

The spacesuit outer layer is Teflon-coated Beta fabric woven of fiberglass strands with a restraint layer, a pressure bladder and an inner high-temperature nylon liner.

Oxygen connection, communications and biomedical data lines attach to fittings on the front of the torso.

Pressure helmet assembly

Feed port

Helmet attaching ring

Zipper access to shoulder disconnect

Penlight pocket

Electrical connector

O_2 inlet

O_2 outlet

Pressure gage

Helmet tie down strap

Wrist disconnect

PGA pressure glove

Protective cover (detached)

Entrance slide fastener flap

Utility pocket

UCT and biomedical injection flap patch

Scissors pocket (detachable)

Check list pocket (detachable)

Data list pocket (detachable)

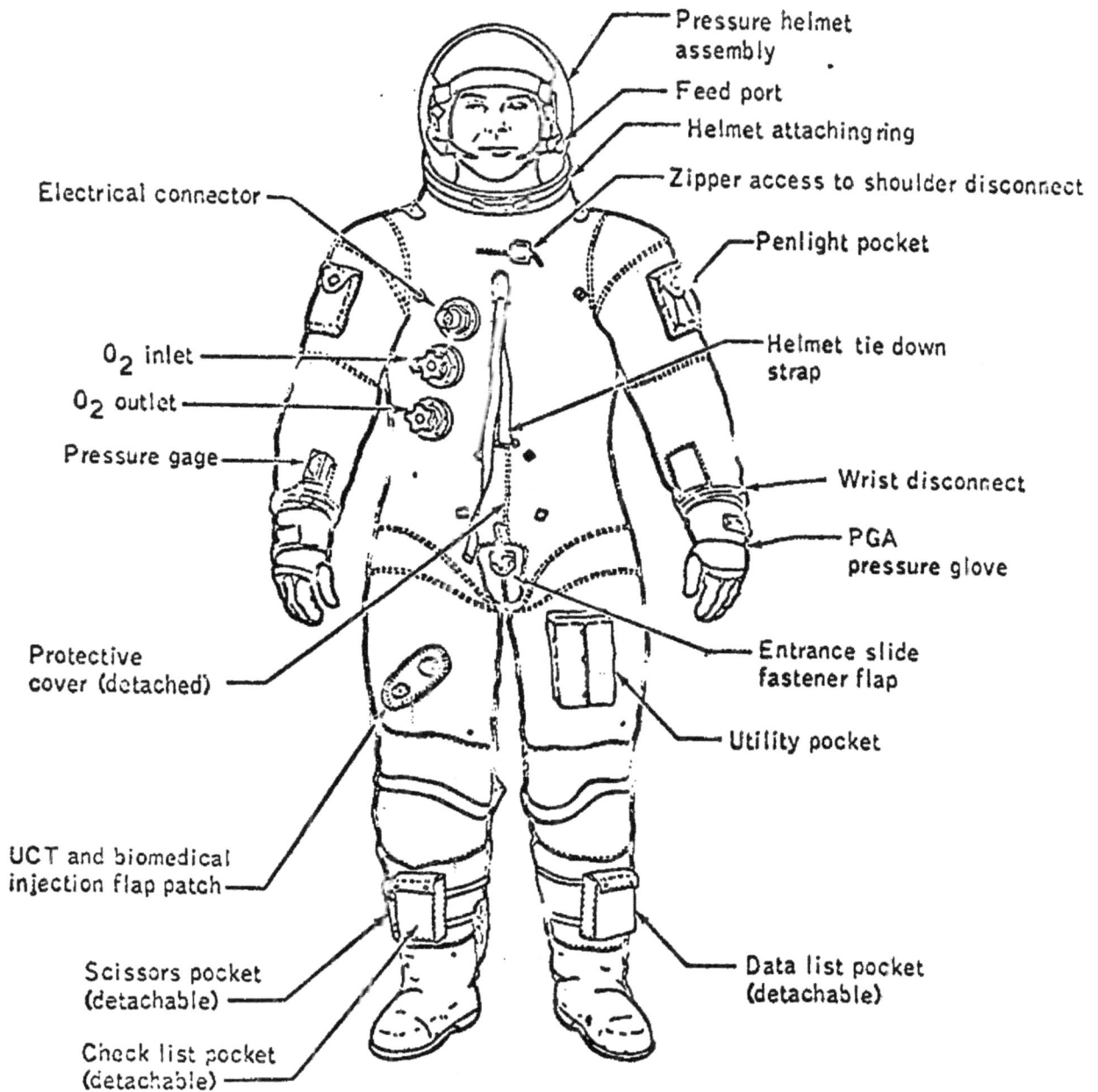

Intravehicular configuration of the PGA.

A one-piece constant wear garment, similar to
"long johns," is worn as an undergarment for the spacesuit
and for the in-flight garment is porous-knit cotton
with a waist-to-neck zipper for donning. Attach points
for the biomedical harness also are provided.

After taking off the spacesuits, the crew will wear
Teflon fabric inflight coveralls over the constant wear
garment. The two-piece coveralls provide warmth in
addition to pockets for personal items. The crew will
wear the inflight coveralls during entry. The soles
of the garment have been fitted with a special metal
heel clip which fits in the couch heel restraint.
Additionally, fitted fluorel foam pads on couch headrests
will provide head restraint during entry. These pads
will be stowed until just prior to entry.

The crewmen will wear communications carriers inside
the pressure helmet. The communications carriers provide
redundancy in that each has two microphones and two
earphones.

A lightweight headset is worn with the inflight
coveralls.

Apollo 8 Crew Meals

The Apollo 8 crew had a wide range of food items
from which to select their daily mission space menu.
More than 60 items comprise the selection list of freeze-
dried bite-size rehydratable foods.

Average daily value of three meals will be 2,500
calories per man.

Unlike Gemini crewmen who prepared their meals
with cold water, Apollo crewmen have running water for
hot meals and cold drinks.

Water is obtained from three sources--a dispenser
for drinking water and two water spigots at the food
preparation station, one supplying water at about 155
degrees F., the other at about 55 degrees F. The
potable water dispenser emits half-ounce spurts with
each squeeze and the food preparation spigots dispense
water in one-ounce increments.

- more -

Spacecraft potable water is supplied from service module fuel cell by-product water.

The day-by-day, meal-by-meal Apollo 8 menu for each crewman is listed on the following page.

Personal Hygiene

Crew personal hygiene equipment aboard Apollo 8 includes body cleanliness items, the waste management system and two medical kits.

Packaged with the food are a toothbrush and a two-ounce tube of toothpaste for each crewman. Each man-meal package contains a 3.5 by 4-inch wet-wipe cleansing towel. Additionally, three packages of 12 by 12-inch dry towels are stowed beneath the command module pilot's couch. Each package contains seven towels. Also stowed under the command module pilot's couch are seven tissue dispensers containing 53 3-ply tissues each.

Solid body wastes are collected in Gemini-type plastic defecation bags which contain a germicide to prevent bacteria and gas formation. The bags are sealed after use and stowed in empty food containers for post-flight analysis.

Urine collection devices are provided for use either while wearing the pressure suit or in the inflight coveralls. The urine is dumped overboard through the spacecraft urine dump valve.

The two medical accessory kits, 6 by 4.5 by 4 inches, are stowed on the spacecraft back wall at the feet of the command module pilot.

The medical kits contain three motion sickness injectors, three pain suppression injectors, one 2-oz bottle first aid ointment, two 1-oz bottle eye drops, three nasal sprays, two compress bandages, 12 adhesive bandages, one oral thermometer and two spare crew biomedical harnesses. Pills in the medical kits are 60 antibiotic, 12 nausea, 12 stimulant, 18 pain killer, 60 decongestant, 24 diarrhea, 72 aspirin and 21 sleeping.

- more -

Apollo 8 (Borman, Lovell, Anders)

	Day 1* 5, and 9	Day 2, 6, and 10	Day 3, 7, and 11	Day 4, 8, and 12

A.	Peaches
	Bacon Squares (8)
	Cinn Tstd Bread Cubes (8)
	Grapefruit Drink
		Canadian Bacon &
		  Applesauce
		Sugar Coated Corn
		  Flakes
		Apricot Cereal Cubes (8)
		Grapefruit Drink
		Orange Drink
			Fruit Cocktail
			Bacon Squares (8)
			Cinn Tstd Bread Cubes (8)
			Cocoa
			Orange Drink
				Canadian Bacon &
				  Applesauce
				Toasted Bread
				  Cubes (8)
				Strawberry Cereal
				  Cubes (6)
				Cocoa
				Orange Drink

B.	Corn Chowder
	Chicken & Gravy
	Toasted Bread Cubes (6)
	Sugar Cookie Cubes (6)
	Cocoa
	Orange Drink
		Tuna Salad
		Chicken & Vegetables
		Cinn Tstd Bread Cubes (8)
		Pineapple Fruitcake (4)
		Pineapple-Grapefruit
		  Drink
			Cream of Chicken Soup
			Beef Pot Roast
			Toasted Bread Cubes (8)
			Butterscotch Pudding
			Grapefruit Drink
				Pea Soup
				Chicken & Gravy
				Cheese Sandwiches
				  (6)
				Bacon Squares
				  (6)
				Grapefruit Drink

C.	Beef & Gravy
	Beef Sandwiches (4)
	Cheese-Cracker Cubes (8)
	Chocolate Pudding
	Orange-Grapefruit Drink
		Spaghetti & Meat Sauce
		Beef Bites (6)
		Bacon Squares (6)
		Banana Pudding
		Grapefruit Drink
			Potato Soup
			Chicken Salad
			Turkey Bites (6)
			Graham Cracker Cubes
			  (6)
			Orange Drink
				Shrimp Cocktail
				Beef Hash
				Cinn Tstd Bread
				  Cubes (8)
				Date Fruitcake
				  (4)
				Orange-Grapefruit
				  Drink

DAYS TOTAL CALORIES	2485	2537	2522	2441

- more -

*Day 1 consists of Meals B and C only; Day 12 consists of Meal A only. Each crewmember will be provided with a total of 33 meals.

Sleep-Work Cycles

At least one crew member will be awake at all times. The normal cycle will be 17 hours of work followed by seven hours of rest. Simultaneous rest periods are scheduled for the command module pilot and the lunar module pilot. When possible, all three crewmen will eat together, with one hour allocated for each meal period.

Sleeping positions in the command module are under the left and right couches, with heads toward the crew hatch. Two lightweight Beta fabric sleeping bags are each supported by two longitudinal straps attaching to lithium hydroxide storage boxes at one end and to the spacecraft pressure vessel inner structure at the other end.

Additional transverse restraint straps have been added to the sleeping bags since the Apollo 7 mission to provide greater sleeping comfort and body restraint in zero-g. The sleeping bags have also been perforated for improved ventilation.

Survival Gear

The survival kit is stowed in two rucksacks in the right-hand forward equipment bay above the lunar module pilot.

Contents of rucksack No. 1 are: two combination survival lights, one desalter kit, three pair sunglasses, one radio beacon, one spare radio beacon battery and spacecraft connector cable, one machete in sheath, three water containers and two containers of Sun lotion. Rucksack No. 2: one three-man life raft with CO_2 inflater, one sea anchor, two sea dye markers, three sunbonnets, one mooring lanyard, three manlines and two attach brackets.

The survival kit is designed to provide a 48-hour postlanding (water or land) survival capability for three crewmen between 40 degrees North and South Latitudes.

Biomedical Inflight Monitoring

The Apollo 8 crew inflight biomedical telemetry data received by the Manned Space Flight Network will be relayed for instantaneous display at Mission Control Center. Heart rate and breathing rate data will be displayed on the flight surgeon's console. Heart rate and respiration rate average, range and deviation are computed and displayed on the digital TV screens.

DETAIL Ⓐ

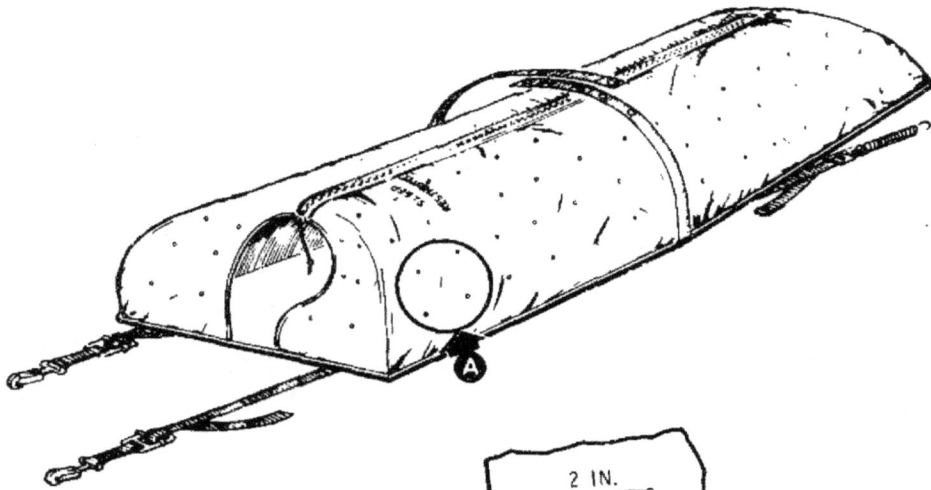

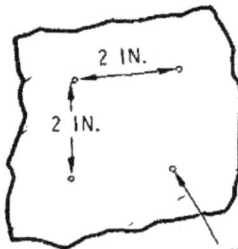

VENTILATION HOLES
.060 IN. DIA

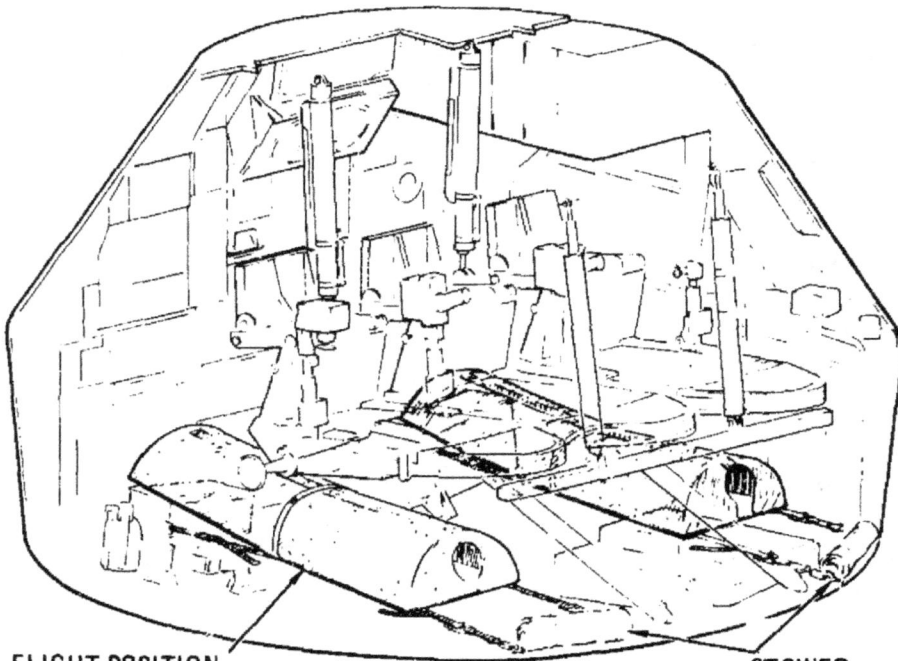

2 IN.

2 IN.

FLIGHT POSITION

STOWED
POSITION

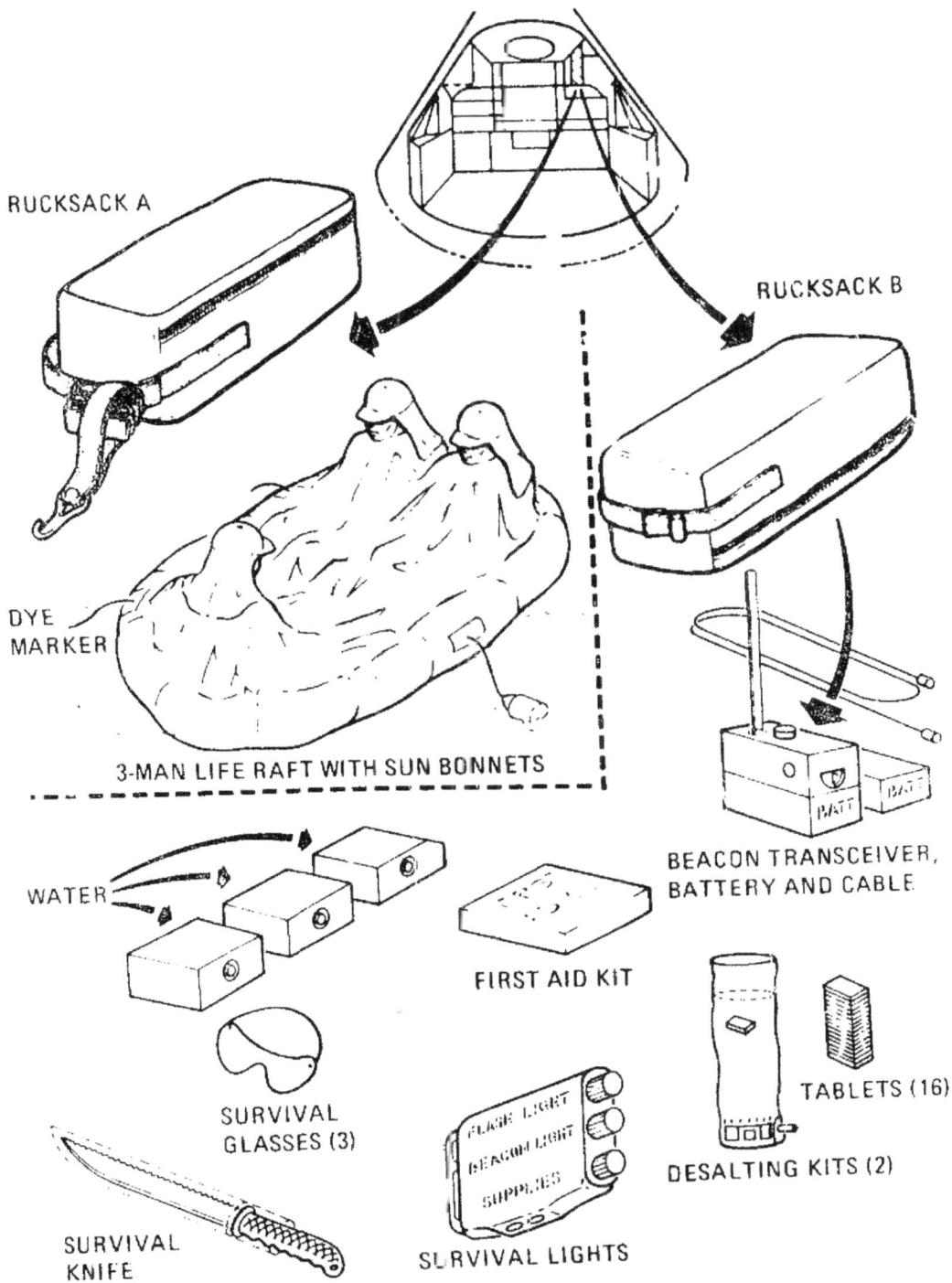

RUCKSACK A

RUCKSACK B

DYE
MARKER

3-MAN LIFE RAFT WITH SUN BONNETS

BEACON TRANSCEIVER,
BATTERY AND CABLE

WATER

FIRST AID KIT

TABLETS (16)

SURVIVAL
GLASSES (3)

DESALTING KITS (2)

SURVIVAL
KNIFE

FLASH LIGHT
BEACON LIGHT
SUPPLIES

SURVIVAL LIGHTS

In addition, the instantaneous heart rate, real time and delayed EKG and respiration are recorded on strip charts for each man.

Biomedical data observed by the flight surgeon and his team in the Life Support Systems Staff Support Room will be correlated with spacecraft and spacesuit environmental data displays.

Blood pressure and body temperature are no longer taken as they were in earlier manned flight programs.

The Crew on Launch Day

Following is a timetable of Apollo 8 crew activities on launch day. (All times are shown in hours and minutes before liftoff.)

T-9:00 - Backup crew alerted

T-8:30 - Backup crew to LC-39A for spacecraft pre-launch checkouts

T-5:00 - Flight crew alerted

T-4:45 - Medical examinations

T-4:15 - Breakfast

T-3:45 - Don pressure suits

T-3:30 - Leave Manned Spacecraft Operations Building for LC-39A via Crew Transfer Van

T-2:30 - Arrive at LC-39A

T-2:37 - Enter elevator to spacecraft level

T-2:40 - Begin spacecraft ingress

Radiation Monitoring

Apollo 8 crew radiation dosages will be closely monitored by onboard dosimeters which either provide crew readouts or telemeter radiation measurements to Manned Space Flight Network stations.

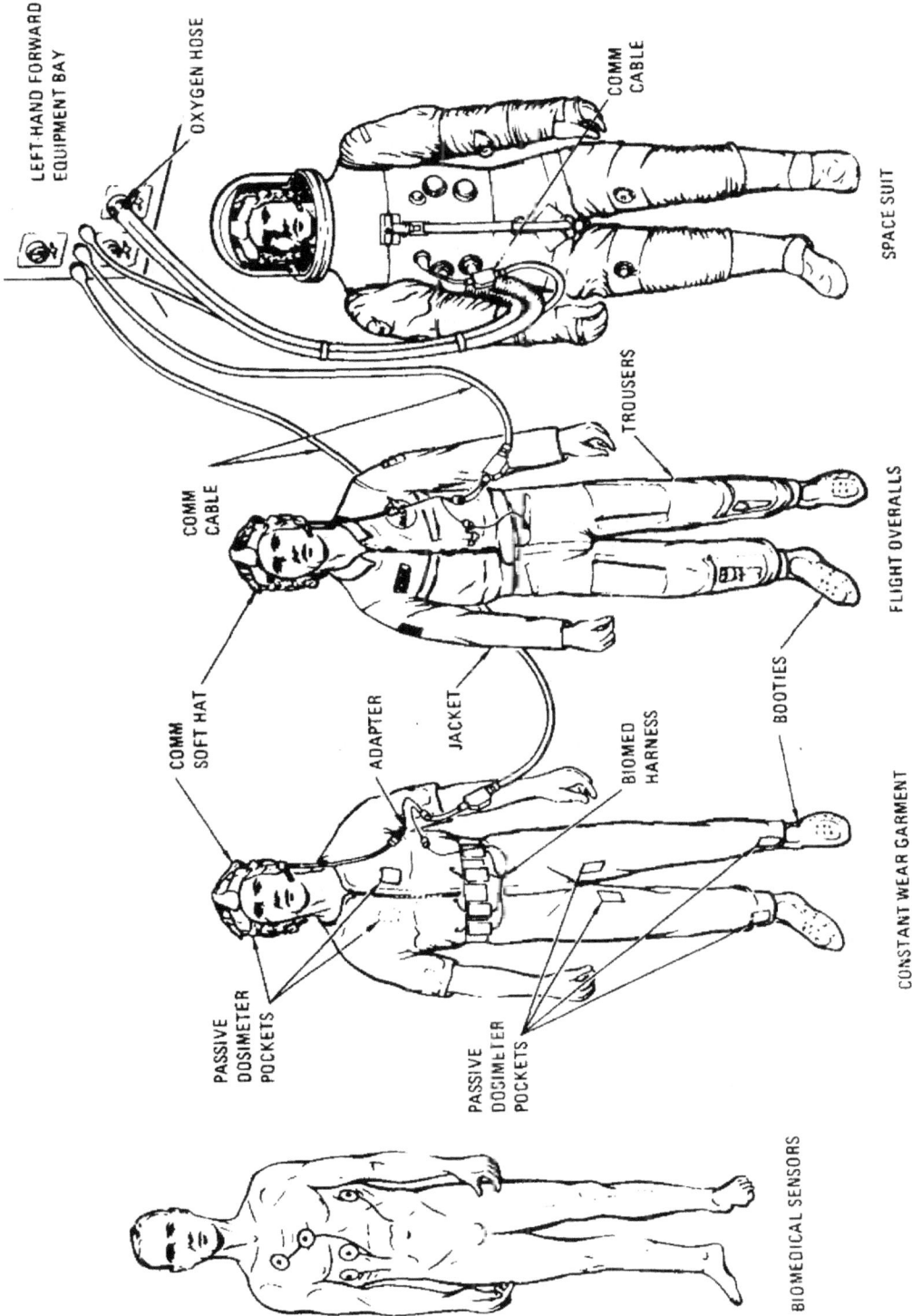

LEFT HAND FORWARD
EQUIPMENT BAY

OXYGEN HOSE

COMM
CABLE

SPACE SUIT

COMM
CABLE

TROUSERS

FLIGHT OVERALLS

COMM
SOFT HAT

ADAPTER

JACKET

BIOMED
HARNESS

BOOTIES

PASSIVE
DOSIMETER
POCKETS

PASSIVE
DOSIMETER
POCKETS

CONSTANT WEAR GARMENT

BIOMEDICAL SENSORS

In addition, Solar Particle Alert Network (SPAN) stations will monitor solar flare activity during the mission to provide forecasts of any increase in radiation.

Five types of radiation measuring devices are carried aboard Apollo 8. Each crewman wears standard passive film dosimeters in the thigh, chest and ankle area which provide cumulative postflight dosage readings. Each man also has a personal radiation dosimeter that can be read for cumulative dosage at any time. They are worn on the right thigh of the pressure garment, and by option on either the shoulder or thigh of the constant wear garment after the pressure suits have been doffed.

Radiation dose rate within the spacecraft cabin is measured by the radiation survey meter, a one-and-a-half pound device mounted in the lower equipment bay near the navigation sextant.

A Van Allen belt dosimeter mounted on the spacecraft girth frame near the lunar module pilot's head measures and telemeters onboard radiation skin dose rates and depth dose rates to network stations.

Proton and alpha particle rates and energies exterior to the spacecraft are measured and telemetered by the nuclear particle detection system mounted on the service module forward bulkhead in the area covered by the fairing around the CM-SM mating line.

SPAN sites keeping tabs on solar flare activity during Apollo 8 will be NASA-operated stations at Manned Spacecraft Center, Carnarvon, Australia, and Canary Islands; and Environmental Sciences Services Administration (ESSA) sites at Boulder, Colo., and Culgoora, Australia.

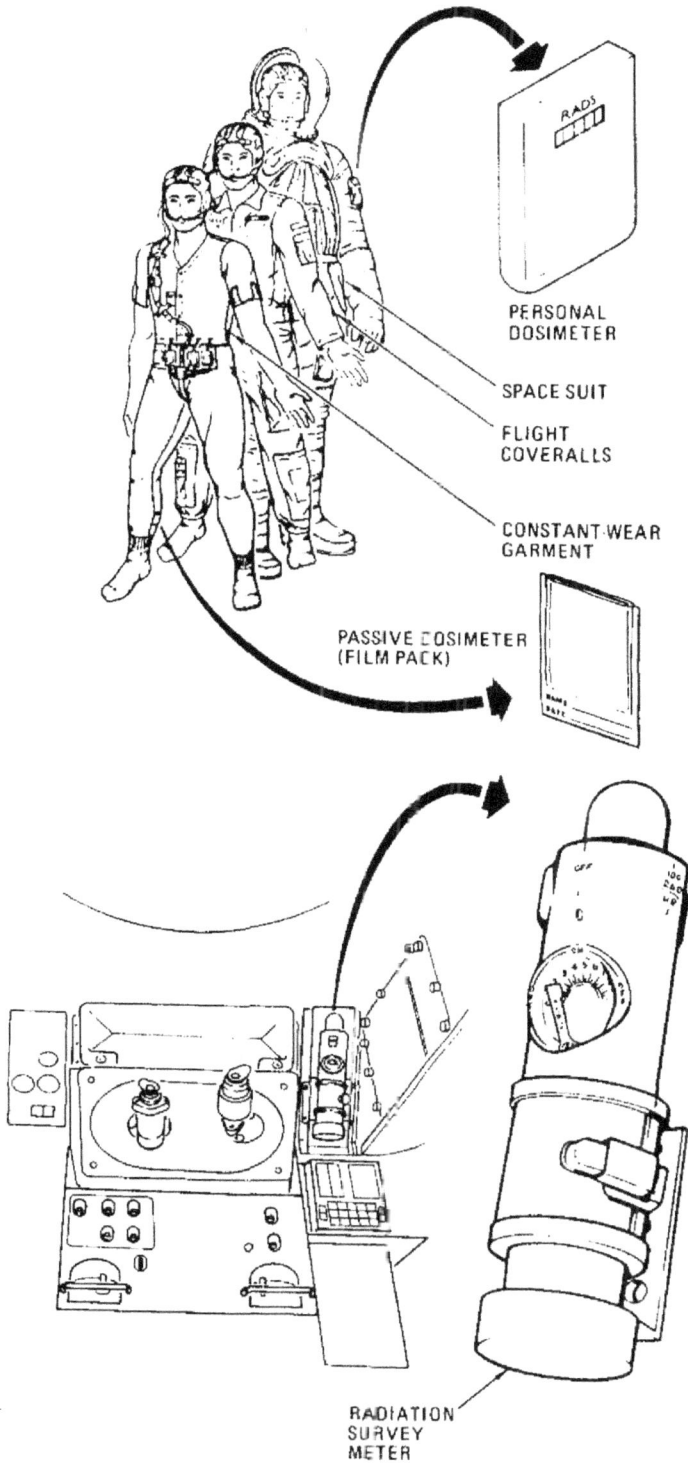

PERSONAL
DOSIMETER

SPACE SUIT

FLIGHT
COVERALLS

CONSTANT WEAR
GARMENT

PASSIVE DOSIMETER
(FILM PACK)

RADIATION
SURVEY
METER

CREW BIOGRAPHIES

NAME: Frank Borman (Colonel, USAF)
      Commander

BIRTHPLACE AND DATE: Born March 14, 1928, in Gary, Ind., but
      grew up in Tucson, Ariz. His parents, Mr. and Mrs. Edwin
      Borman, now reside in Phoenix, Ariz.

PHYSICAL DESCRIPTION: Blond hair; blue eyes; height: 5 feet
      10 inches; weight: 163 pounds.

EDUCATION: Received a Bahcelor of Science degree from the
      United States Military Academy at West Point in 1950
      and a Master of Science degree in Aeronautical Engineering
      from the California Institute of Technology, Pasadena,
      Calif., in 1957.

MARITAL STATUS: Married to the former Susan Bugbee of Tucson,
      Ariz.; her mother, Mrs. Ruth Bugbee, resides in Tucson,
      Ariz.

CHILDREN: Fredrick, October 4, 1951; Edwin, July 20, 1953.

OTHER ACTIVITIES: He enjoys hunting and water skiing.

ORGANIZATIONS: Member of the American Institute of Aeronautics
      and Astronautics and the Society of Experimental Test
      Pilots.

SPECIAL HONORS: Awarded the NASA Exceptional Service Medal,
      Air Force Astronaut Wings, and Air Force Distinguished
      Flying Cross; recipient of the 1966 American Astro-
      nautical Flight Achievement Award and the 1966 Air Force
      Association David C. Schilling Flight Trophy; co-
      recipient of the 1966 Harmon International Aviation
      Trophy; and recipient of the California Institute of
      Technology Distinguished Alumni Service Award for 1966.

EXPERIENCE: Borman, an Air Force Colonel, entered the Air
      Force after graduation from West Point and received
      his pilot training at Williams Air Force Base, Arizona.

      From 1951 to 1956, he was assigned to various fighter
      squadrons in the United States and the Philippines..
      He became an instructor of thermo-dynamics and fluid
      mechanics at the Military Academy in 1957 and subsequently
      attended the USAF Aerospace Research Pilots School from
      which he graduated in 1960. He remained there as an
      instructor until 1962.

He has accumulated over 5,500 hours flying time,
including 4,500 hours in jet aircraft.

CURRENT ASSIGNMENT: Colonel Borman was selected as an
astronaut by NASA in September 1962. He has performed
a variety of special duties, including an assignment
as backup command pilot for the Gemini 4 flight and as
a member of the Apollo 204 Review Board.

As command pilot of the history-making Gemini 7 mission,
launched on Dec. 4, 1965, he participated in estab-
lishing a number of space "firsts"--among which are the
longest manned space flight (330 hours and 35 minutes)
and the first rendezvous of two manned maneuverable
spacecraft as Gemini 7 was joined in orbit by Gemini 6.

-end-

NAME: James A. Lovell, Jr. (Captain, USN)
       Command Module Pilot

BIRTHPLACE AND DATE: Born March 25, 1928, in Cleveland, Ohio.
     His mother, Mrs. Blanche Lovell, resides at Edgewater
     Beach, Fla.

PHYSICAL DESCRIPTION: Blond hair; blue eyes; height: 5 feet
     11 inches; weight: 170 pounds.

EDUCATION: Graduated from Juneau High School, Milwaukee,
     Wisc.; attended the University of Wisconsin for 2
     years, then received a Bachelor of Science degree from
     the United States Naval Academy in 1952.

MARITAL STATUS: Married to the former Marilyn Gerlach of
     Milwaukee, Wisc. Her parents, Mr. and Mrs. Carl
     Gerlach, are residents of Milwaukee.

CHILDREN: Barbara L., October 13, 1953; James A., February
     15, 1955; Susan K., July 14, 1958; Jeffrey C., January
     14, 1966.

OTHER ACTIVITIES: His hobbies are golf, swimming, handball,
     and tennis.

ORGANIZATIONS: Member of the Society of Experimental Test
     Pilots and the Explorers Club.

SPECIAL HONORS: Awarded two NASA Exceptional Service Medals,
     the Navy Astronaut Wings, two Navy Distinguished Flying
     Crosses, and the 1967 FAI Gold Space Medal (Athens, Greece);
     and co-recipient of the 1966 American Astronautical
     Society Flight Achievement Award and the Harmon Inter-
     national Aviation Trophy in 1966 and 1967.

EXPERIENCE: Lovell, a Navy Captain, received flight training
     following graduation from Annapolis.

     He has had numerous naval aviator assignments including
     a 4-year tour as a test pilot at the Naval Air Test
     Center, Patuxent River, Md. While there he served as
     program manager for the F4H weapon system evaluation.
     A graduate of the Aviation Safety School of the University
     of Southern California, he also served as a flight
     instructor and safety officer with Fighter Squadron 101
     at the Naval Air Station, Oceana, Va.

     Of the 4,000 hours flying time he has accumulated,
     more than 3,000 hours are in jet aircraft.

-more-

CURRENT ASSIGNMENT: Captain Lovell was selected as an
astronaut by NASA in September 1962. He has since
served as backup pilot for the Gemini 4 flight and
backup command pilot for the Gemini 9 flight.

On Dec. 4, 1965, he and command pilot Frank Borman
were launched into space on the history-making Gemini
7 mission. The flight lasted 330 hours and 35 minutes,
during which the following space "firsts" were accomplished: longest manned space flight; first rendezvous
of two manned maneuverable spacecraft, as Gemini 7 was
joined in orbit by Gemini 6; and longest multi-manned
space flight. It was also on this flight that numerous
technical and medical experiments were completed suc-
cessfully.

The Gemini 12 mission, with Lovell and pilot Edwin
Aldrin, began on Nov. 11, 1966. This 4-day 59-revolu-
tion flight brought the Gemini Program to a successful
close. Major accomplishments of the 94-hour 35-minute
flight included a third-revolution rendezvous with the
previously launched Agena (using for the first time
backup onboard computations due to a radar failure);
a tethered station-keeping exercise; retrieval of a
micro-meteorite experiment package from the spacecraft
exterior; an evaluation of the use of body restraints
specially designed for completing work tasks outside of
the spacecraft; and completion of numerous photographic
experiments, the highlights of which are the first pictures
taken from space of an eclipse of the Sun.

Gemini 12 ended when retrofire occurred at the
beginning of the 60th revolution, followed by the
second consecutive fully automatic controlled reentry
of a spacecraft, and a landing in the Atlantic within $2\frac{1}{2}$
miles of the prime recovery ship USS WASP.

As a result of his participation in this flight,
Lovell holds the space endurance record, with 425 hours
and 10 minutes, for total time spent in space. Aldrin
established a new EVA record by completing $5\frac{1}{2}$ hours
outside the spacecraft during two standup EVAs and one
umbilical EVA.

SPECIAL ASSIGNMENT: In addition to his regular duties as a
member of the astronaut group, Captain Lovell was
selected in June 1967 to serve as Special Consultant
to the President's Council on Physical Fitness.

-end-

NAME: William A. Anders (Major, USAF) , Lunar Module Pilot

BIRTHPLACE AND DATE: Born October 17, 1933, in Hong Kong; his parents, Commander (USN retired) and Mrs. Arthur F. Anders, now reside in La Mesa, Calif.

PHYSICAL DESCRIPTION: Brown hair; blue eyes; height: 5 feet 8 inches; weight: 145 pounds.

EDUCATION: Received a Bachelor of Science degree from the United States Naval Academy in 1955 and a Master of Science degree in Nuclear Engineering from the Air Force Institute of Technology at Wright-Patterson Air Force Base, Ohio, in 1962.

MARITAL STATUS: Married to the former Valerie E. Hoard of Lemon Grove, Calif., daughter of Mr. and Mrs. Henry G. Hoard, of Oceanside, Calif.

CHILDREN: Alan, February 1957; Glen, July 1958; Gayle, December 1960; Gregory, December 1962; Eric, July 1964.

OTHER ACTIVITIES: His hobbies are fishing, flying, camping, and water skiing; and he also enjoys soccer.

ORGANIZATIONS: Member of the American Nuclear Society and Tau Beta Pi.

SPECIAL HONORS: Awarded the Air Force Commendation Medal.

EXPERIENCE: Anders, an Air Force Major, was commissioned in the Air Force upon graduation from the Naval Academy. After Air Force flight training, he served as a fighter pilot in all-weather interceptor squadrons of the Air Defense Command.

After his graduate training, he served as a nuclear engineer and instructor pilot at the Air Force Weapons Laboratory, Kirtland Air Force Base, N.M., where he was responsible for technical management of radiation nuclear power reactor shielding and radiation effects programs.

He has logged more than 3,000 hours flying time.

CURRENT ASSIGNMENT: Major Anders was one of the third group of astronauts selected by NASA in October 1963. He has since served as backup pilot for the Gemini 11 mission.

-end-

LUNAR DESCRIPTION

Terrain - Mountainous and crater-pitted, the former rising thousands of feet and the latter ranging from a few inches to 180 miles in diameter. The craters are thought to be formed by the impact of meteorites. The surface is covered with a layer of fine-grained material resembling silt or sand, as well as small rocks.

Environment - No air, no wind, and no moisture. The temperature ranges from 250 degrees in the two-week lunar day to 280 degrees below zero in the two-week lunar night. Gravity is one-sixth that of Earth. Micrometeoroids pelt the Moon (there is no atmosphere to burn them up). Radiation might present a problem during periods of unusual solar activity.

Dark Side - The dark or hidden side of the Moon no longer is a complete mystery. It was first photographed by a Russian craft and since then has been photographed many times, particularly by NASA's Lunar Orbiter spacecraft.

Origin - There is still no agreement among scientists on the origin of the Moon. The three theories: (1) the Moon once was part of Earth and split off into its own orbit, (2) it evolved as a separate body at the same time as Earth, and (3) it formed elsewhere in space and wandered until it was captured by Earth's gravitational field.

Earth to Moon Distances

Date	At	Surface to Surface
Dec. 21	5 p.m. EST	220,074 statute
Dec. 22	6 p.m. EST	223,337 statute
Dec. 23	7 p.m. EST	227,182 statute
Dec. 24	7:30 p.m. EST	231,238 statute
Dec. 25	8 p.m. EST	235,186 statute
Dec. 26	9 p.m. EST	238,751 statute
Dec. 27	10 p.m. EST	241,779 statute

-more-

Physical Facts

Diameter	2,160 miles (about ¼ that of Earth)
Circumference	6,790 miles (about ¼ that of Earth)
Distance from Earth	238,857 miles (mean; 221,463 minimum to 252,710 maximum)
Surface temperature	250 (Sun at zenith)-280 (night)
Surface gravity	1/6 that of Earth
Mass	1/100th that of Earth
Volume	1/50th that of Earth
Lunar day and night	14 Earth days each
Mean velocity in orbit	2,287 miles per hour
Escape velocity	1.48 miles per second
Month (period of rotation around Earth)	27 days, 7 hours, 43 minutes

-more-

APOLLO PROGRAM MANAGEMENT/CONTRACTORS

Direction of the Apollo Program, the United States' effort to land men on the Moon and return them safely to Earth before 1970, is the responsibility of the Office of Manned Space Flight (OMSF), National Aeronautics and Space Administration, Washington, D.C.

NASA Manned Spacecraft Center (MSC), Houston, is responsible for development of the Apollo spacecraft, flight crew training and flight control.

NASA Marshall Space Flight Center (MSFC), Huntsville, Ala., is responsible for development of the Saturn launch vehicles.

NASA John F. Kennedy Space Center (KSC), Fla., is responsible for Apollo/Saturn launch operations.

NASA Goddard Space Flight Center (GSFC), Greenbelt, Md., manages the Manned Space Flight Network under the direction of the NASA Office of Tracking and Data Acquisition (OTDA).

Apollo/Saturn Officials

Dr. George E. Mueller	Associate Administrator for Manned Space Flight, NASA Headquarters
Maj. Gen. Samuel C. Phillips	Director, Apollo Program Office, OMSF, NASA Headquarters
George H. Hage	Deputy Director, Apollo Program Office, OMSF, NASA Headquarters
William C. Schneider	Apollo Mission Director, OMSF, NASA Headquarters
Chester M. Lee	Assistant Mission Director, OMSF, NASA Headquarters

- more -

Col. Thomas H. McMullen   Assistant Mission Director,
                                 OMSF, NASA Headquarters

Dr. Robert R. Gilruth   Director, Manned Spacecraft
                                 Center, Houston

George M. Low   Manager, Apollo Spacecraft
                                 Program, MSC

Kenneth S. Kleinknecht   Manager, Command and Service
                                 Modules, Apollo Spacecraft
                                 Program Office, MSC

Donald K. Slayton   Director, Flight Crew Opera-
                                 tions, MSC

Christopher C. Kraft, Jr.   Director Flight Operations,
                                 MSC

Clifford E. Charlesworth   Apollo 8 Flight Directors,
Glynn S. Lunney   Flight Operations, MSC
M.L. Windler

Dr. Wernher von Braun   Director, Marshall Space
                                 Flight Center, Huntsville, Ala.

Brig. Gen. Edmund F. O'Connor   Director, Industrial
                                 Operations, MSFC

Lee B. James   Manager, Saturn V Program
                                 Office, MSFC

William D. Brown   Manager, Engine Program
                                 Office, MSFC

Dr. Kurt H. Debus   Director, John F. Kennedy
                                 Space Center, Fla.

Miles Ross   Deputy Director, Center
                                 Operations, KSC

Rocco A. Petrone   Director, Launch Operations,
                                 KSC

Walter J. Kapryan   Deputy Director, Launch
                                 Operations, KSC

Dr. Hans F. Gruene   Director, Launch Vehicle
                                 Operations, KSC

- more -

Rear Adm. Roderick O. Middleton   Manager, Apollo Program
Office, KSC

John J. Williams   Director, Spacecraft
Operations, KSC

Paul C. Donnelly   Launch Operations Manager, KSC

Gerald M. Truszynski   Associate Administrator,
Tracking and Data Acquisition,
NASA Headquarters

H. R. Brockett   Deputy Associate Administrator,
OTDA, NASA Headquarters

Norman Pozinsky   Director, Network Support
Implementation Division, OTDA

Dr. John F. Clark   Director, Goddard Space
Flight, Greenbelt, Md.

Ozro M. Covington   Assistant Director for Manned
Space Flight Tracking, GSFC

Henry F. Thompson   Deputy Assistant Director
for Manned Space Flight
Support, GSFC

H. William Wood   Chief, Manned Flight Operations
Division, GSFC

Tecwyn Roberts   Chief, Manned Flight Engineering
Division, GSFC

L. R. Stelter   Chief, NASA Communications
Division, GSFC

Maj. Gen. Vincent G. Huston   USAF, DOD Manager of Manned
Space Flight Support Operations

Maj. Gen. David M. Jones   USAF, Deputy DOD Manager of
Manned Space Flight Support
Operations, Commander USAF
Eastern Test Range

Rear Adm. F. E. Bakutis   USN, Commander Combined
Task Force 130 Pacific
Recovery Area (Primary)

The Boeing Co. New Orleans	First Stages (SIC) of Saturn V Flight Vehicles, Saturn V Systems Engineering and Integration Ground Support Equipment
North American Rockwell Corp. Space Division Seal Beach, Calif.	Development and Production of Saturn V Second Stage (S-II)
McDonnell Douglas Astronautics Co. Huntington Beach, Calif.	Development and Production of Saturn V Third Stage (S-IVB)
International Business Machines Federal Systems Division Huntsville, Ala.	Instrument Unit (Prime Contractor)
Bendix Corp., Navigation and Control Div. Teterboro, N.J.	Guidance Components for Instrument Unit (Including ST-124M Stabilized Platform)
Trans World Airlines, Inc.	Installation Support, KSC
Federal Electric Corp.	Communications and Instrumentation Support, KSC
Bendix Field Engineering Corp.	Launch Operations/Complex Support, KSC
Catalytic-Dow	Facilities Engineering and Modifications, KSC
ILC Industries Dover, Del.	Space Suits
Radio Corporation of America Van Nuys, Calif.	110A Computer - Saturn Checkout
Sanders Associates Nashua, New Hampshire	Operational Display Systems Saturn
Brown Engineering Huntsville, Alabama	Discrete Controls
Ingalls Iron Works Birmingham, Alabama	Mobile Launchers (structural work)

- more -

Rear Adm. P. S. McManus   USN, Commander Combined
                                 Task Force 140 Atlantic
                                 Recovery Area

Col. Royce G. Olson   USAF, Director, DOD Manned
                                 Space Flight Office

Brig. Gen. Allison C. Brooks   USAF, Commander Aerospace
                                 Rescue and Recovery Service

Major Apollo/Saturn V Contractors

Contractor   Item

Bellcomm   Apollo Systems Engineering
Washington, D.C.

The Boeing Co.   Technical Integration and
Washington, D.C.   Evaluation

General Electric-Apollo   Apollo Checkout and
Support Department,   Reliability
Daytona Beach, Fla.

North American Rockwell Corp.   Spacecraft Command and
Space Division, Downey, Calif.   Service Modules

Grumman Aircraft Engineering   Lunar Module
   Corp.,
Bethpage, N.Y.

Massachusetts Institute of   Guidance & Navigation
Technology, Cambridge, Mass.   (Technical Management)

General Motors Corp., AC   Guidance & Navigation
Electronics Division,   (Manufacturing)
Milwaukee

TRW Systems Inc.   Trajectory Analysis
Redondo Beach, Calif.

Avco Corp., Space Systems   Heat Shield Ablative
Division, Lowell, Mass.   Material

North American Rockwell Corp.   J-2 Engines, F-1 Engines
Rocketdyne Division
Canoga Park, Calif.

- more -

Smith/Ernst (Joint Venture) Tampa, Florida Washington, D.C.	Electrical Mechanical Portion of MLs
Power Shovel, Inc. Marion, Ohio	Crawler-Transporter
Hayes International Birmingham, Alabama	Swing Arm

APOLLO 8 GLOSSARY

Ablating Materials--Special heat-dissipating materials on the surface of a spacecraft that can be sacrificed (carried away, vaporized) during re-entry.

Abort--The cutting short of an aerospace mission before it has accomplished its objective.

Accelerometer--An instrument to sense accelerative forces and convert them into corresponding electrical quantities usually for controlling, measuring, indicating or recording purposes.

Adapter Skirt--A flange or extension of a stage or section that provides a ready means of fitting another stage or section to it.

Antipode--Point on surface of planet exactly 180 degrees opposite from reciprocal point on a line projected through center of body. In Apollo 8 usage, antipode refers to a line from the center of the Moon through the center of the Earth and projected to the Earth surface on the opposite side. The antipode crosses the mid-Pacific recovery line along the 165th meridian of longitude once each 24 hours.

Apocynthion--Point at which object in lunar orbit is farthest from lunar surface--object having been launched from body other than Moon. (Cynthia, Roman goddess of Moon).

Apogee--The point at which a moon or artificial satellite in its orbit is farthest from Earth.

Apolune--Point at which object launched from the Moon into lunar orbit is farthest from lunar surface. e.g. Ascent stage of lunar module after staging into lunar orbit following lunar landing.

Attitude--The position of an aerospace vehicle as determined by the inclination of its axes to some frame of reference; for Apollo, an inertial, space-fixed reference is used.

Burnout--The point when combustion ceases in a rocket engine.

Canard--A short, stubby wing-like element affixed to an aircraft or spacecraft to provide better stability.

Celestial Guidance--The guidance of a vehicle by reference to celestial bodies.

Celestial Mechanics--The science that deals primarily with the effect of force as an agent in determining the orbital paths of celestial bodies.

Cislunar--Adjective referring to space between Earth and the Moon, or between Earth and Moon's orbit.

Closed Loop--Automatic control units linked together with a process to form an endless chain.

Control System--A system that serves to maintain attitude stability during forward flight and to correct deflections.

Deboost--A retrograde maneuver which lowers either perigee or apogee of an orbiting spacecraft. Not to be confused with deorbit.

Delta V--Velocity change.

Digital Computer--A computer in which quantities are represented numerically and which can be used to solve complex problems.

Down-Link--The part of a communication system that receives, processes and displays data from a spacecraft.

Entry Corridor--The final flight path of the spacecraft before and during Earth re-entry.

Escape Velocity--The speed a body must attain to overcome a gravitational field, such as that of Earth; the velocity of escape at the Earth's surface is 36,700 feet-per-second.

Explosive Bolts--Bolts surrounded with an explosive charge which can be activated by an electrical impulse.

Fairing--A piece, part or structure having a smooth, streamlined outline, used to cover a nonstreamlined object or to smooth a junction.

Fuel Cell--An electrochemical generator in which the chemical energy from the reaction of oxygen and a fuel is converted directly into electricity.

G or G Force--Force exerted upon an object by gravity or by reaction to acceleration or deceleration, as in a change of direction: one G is the measure of the gravitational pull required to move a body at the rate of about 32.16 feet-per-second.

Gimballed Motor--A rocket motor mounted on gimbal; i.e., on a contrivance having two mutually perpendicular axes of rotation, so as to obtain pitching and yawing correction moments.

Guidance System--A system which measures and evaluates flight information, correlates this with target data, converts the result into the conditions necessary to achieve the desired flight path, and communicates this data in the form of commands to the flight control system.

Heliocentric--Sun-centered orbit or other activity which has the Sun as its center.

Inertial Guidance--A sophisticated automatic navigation system using gyroscopic devices, etc., for high-speed vehicles. It absorbs and interprets such data as speed, position, etc., and automatically adjusts the vehicle to a predetermined flight path. Essentially, it knows where it's going and where it is by knowing where it came from and how it got there. It does not give out any signal so it cannot be detected by radar or jammed.

Injection--The process of injecting a spacecraft into a calculated orbit.

Multiplexing--The simultaneous transmission of two or more signals within a single channel. The three basic methods of multiplexing involve the separation of signals by time division, frequency division and phase division.

Optical Navigation--Navigation by sight, as opposed to mathematical methods.

Oxidizer--In a rocket propellant, a substance such as liquid oxygen or nitric acid that yields oxygen for burning the fuel.

Penumbra--Semi-dark portion of a shadow in which light is partly cut off. e.g., Surface of Moon or Earth away from Sun. (See umbra.)

Pericynthion--Point nearest moon of object in lunar orbit--object having been launched from body other than moon.

Perigee--The point at which a moon or an artificial satellite in its orbit is closest to the Earth.

Perilune--The point at which a satellite (e.g., a spacecraft) in its orbit is closest to the Moon: differs from pericynthion in that the orbit is Moon-originated.

Pitch--The movement of a space vehicle about an axis (Y) that is perpendicular to its longitudinal axis.

Re-entry--The return of a spacecraft that re-enters the atmosphere after flight above it.

Retrorocket--A rocket that gives thrust in a direction opposite to the direction of the object's motion.

Roll--The movements of a space vehicle about its longitudinal (X) axis.

S Band--A radio-frequency band of 1550 to 5200 megacycles per second.

Selenographic--Adjective relating to physical geography of Moon. Specifically, positions on lunar surface as measured in latitude from lunar equator and in longitude from a reference lunar meridian.

Selenocentric--Adjective referring to orbit having Moon as center. (Selene, Gr. moon)

Sidereal--Adjective relating to measurement of time, position or angle in relation to the celestial sphere and the vernal equinox.

Telemetering--A system for taking measurements within an aerospace vehicle in flight and transmitting them by radio to a ground station.

Terminator--Separation line between lighted and dark portions of celestial body which is not self luminous.

Ullage--The volume in a closed tank or container above the surface of a stored liquid. Also the ratio of this volume to the total volume of the tank.

Umbra--Darkest part of a shadow in which light is completely absent. e.g., Surface of Moon or Earth away from Sun.

Up-Link Data--Telemetry information from the ground.

Yaw--Displacement of a space vehicle from its vertical (Z) axis.

-more-

Apollo 8 Acronyms

(Note: This list makes no attempt to include all
Apollo program acronyms. Listed are several
acronyms that are encountered for the first time
in the Apollo 8 mission.)

AK	Apogee kick
COI	Contingency orbit insertion
EOI	Earth orbit insertion
HGA	High-gain antenna
IRIG	Inertial reference integrating gyro
LOI	Lunar orbit insertion
LPO	Lunar parking orbit
LTAB	Lunar (module) test article B
MCC	Midcourse correction, Mission Control Center
MSI	Moon sphere of influence
REFSMMAT	Reference to stable member matrix
TEI	Transearth injection
TEMCC	Transearth midcourse correction
TLI	Translunar injection
TLMCC	Translunar midcourse correction

-more-

Conversion Factors

	Multiply	By	To Obtain
Distance:			
	feet	0.3048	meters
	meters	3.281	feet
	kilometers	3281	feet
	statute miles	1.609	kilometers
	nautical miles	1.852	kilometers
	nautical miles	1.1508	statute miles
	statute miles	0.86898	nautical miles
Velocity:			
	feet/sec	0.3048	meters/sec
	meters/sec	3.281	feet/sec
	feet/sec	0.6818	statute miles/hr
	statute miles/hr	1.609	km/hr
	km/hr	0.6214	statute miles/hr
Liquid measure, weight:			
	gallons	3.785	liters
	liters	0.2642	gallons
	pounds	0.4536	kilograms
	kilograms	2.205	pounds
	pounds	14.0	stones

-more-

Electrical Power Conversion

Voltage X Current in amps = power in watts

Watts ÷ voltage = amps

Watts ÷ amps = voltage

-end-

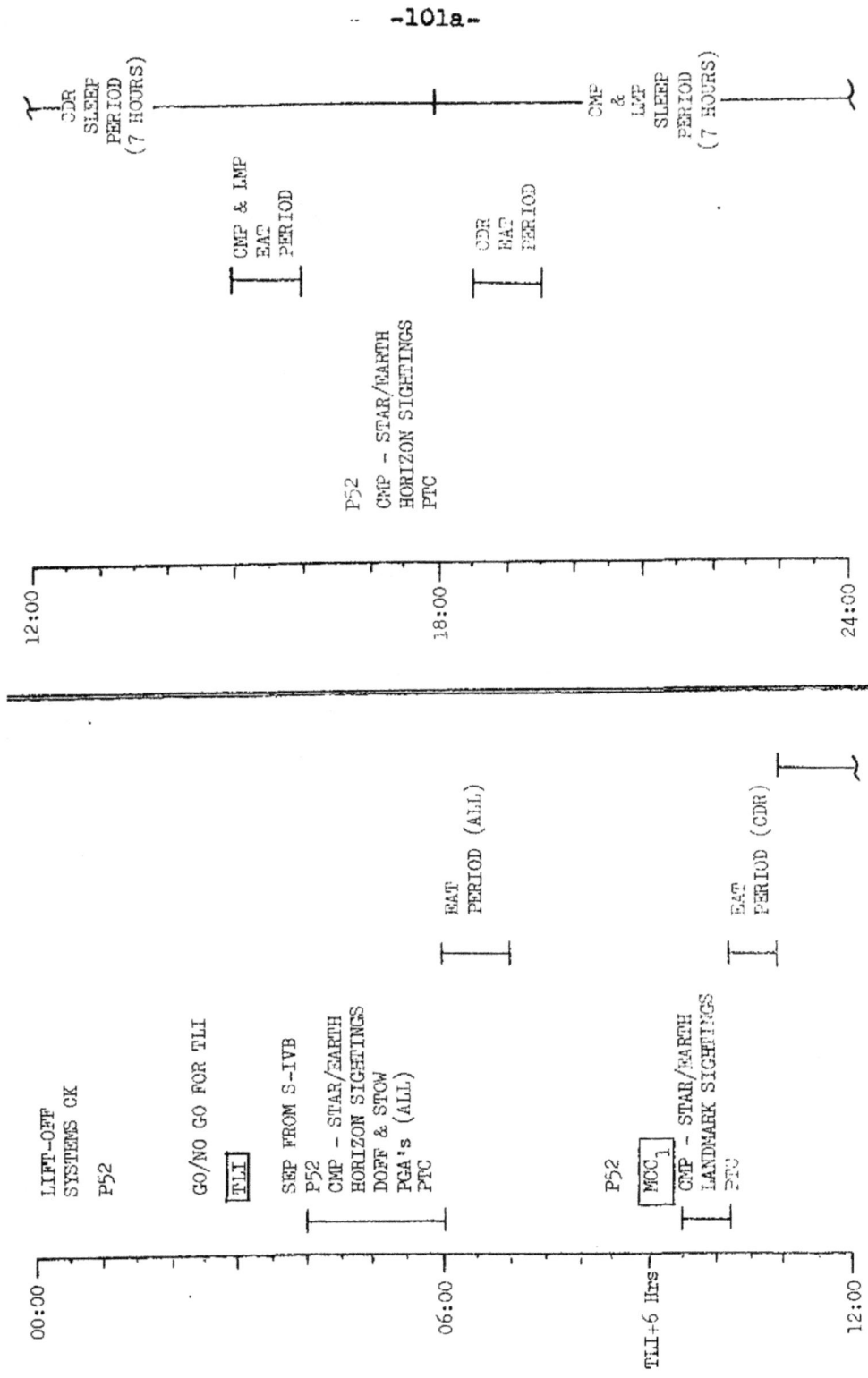

FLIGHT PLAN

00:00

LIFT-OFF
SYSTEMS CK

P52

GO/NO GO FOR TLI

TLI

SEP FROM S-IVB
P52
CMP - STAR/EARTH
HORIZON SIGHTINGS
DOFF & STOW
PGA's (ALL)
PTC

06:00

EAT
PERIOD (ALL)

P52

MCC₁
CMP - STAR/EARTH
LANDMARK SIGHTINGS
PTC

EAT
PERIOD (CDR)

TLI+6 Hrs

12:00

12:00

CDR
SLEEP
PERIOD
(7 HOURS)

CMP & LMP
EAT
PERIOD

18:00

P52
CMP - STAR/EARTH
HORIZON SIGHTINGS
PTC

CDR
EAT
PERIOD

CMP
&
LMP
SLEEP
PERIOD
(7 HOURS)

24:00

MISSION	EDITION	DATE	TIME	DAY/REV	PAGE
AS503/103	FINAL	November 22, 1966	00:00 -24:00	1	5-1

FLIGHT PLANNING BRANCH

MSC FORM 1186 (SEP 67)

FLIGHT PLAN

CMP & LMP SLEEP PERIOD (7 HOURS)

CDR EAT PERIOD

EAT PERIOD (ALL)

P52
CMP – STAR/LUNAR HORIZON SIGHTINGS

MCC₃
CMP – STAR/EARTH HORIZON SIGHTINGS

36:00

42:00

LOI -22 Hr

48:00

EAT PERIOD (ALL)

CMP & LMP EAT PERIOD

CDR SLEEP PERIOD (7 HOURS)

CMP & LMP EAT PERIOD

P52
CMP – STAR/EARTH HORIZON SIGHTINGS
PTC

MCC₂
CMP – STAR/EARTH HORIZON SIGHTINGS
PTC

TV

P52
CMP – STAR/EARTH HORIZON SIGHTINGS
PTC

24:00

TLI + 25 Hrs

30:00

36:00

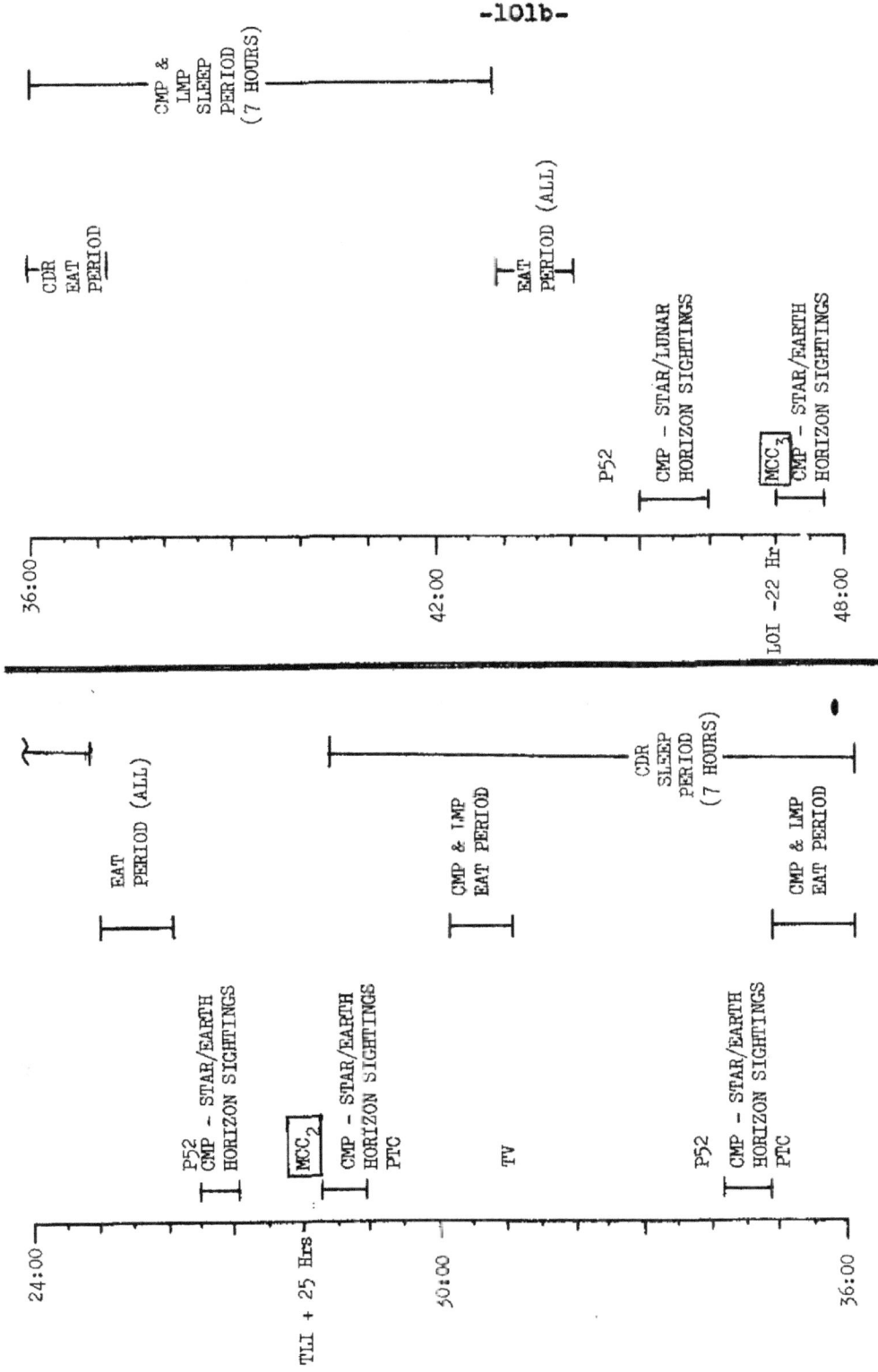

MISSION	EDITION	DATE	DAY/REV	TIME	PAGE
AS503/103	FINAL	November 22, 1968	2	24:00 – 48:00	5-2

FLIGHT PLANNING BRANCH

MSC FORM 1106 (SEP 67)

FLIGHT PLAN

EAT PERIOD (ALL)

PTC

P52 - STAR/LUNAR HORIZON SIGHTINGS

PTC

TV

CMP & LMP EAT PERIOD

CDR EAT PERIOD

48:00

54:00

60:00

P52 MCC₄

PTC

LOI -8 Hrs

60:00

CDR SLEEP PERIOD (7 HOURS)

P52
PTC

LOI₁ ATT CK
PTC

66:00

Rev 1

CMP & LMP SLEEP PERIOD (6 HOURS)

LOI ATT CK

P52 LOI₁

EAT PERIOD (ALL)

LMK FAMILIARIZATION AND TV

Rev 2
72:00

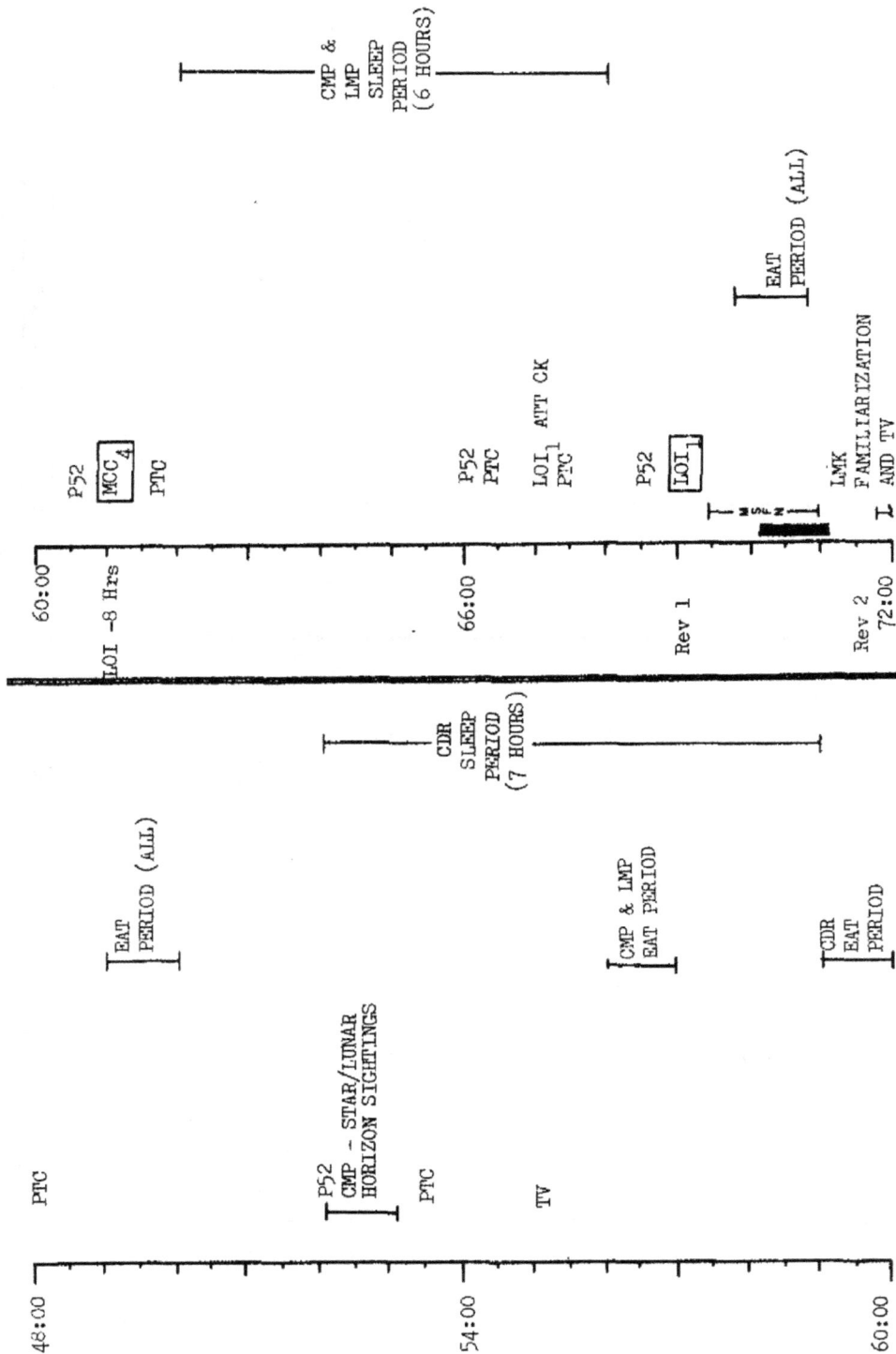

MISSION	EDITION	DATE	TIME	DAY/REV	PAGE
AS503/103	FINAL	November 22, 1968	48:00 - 72:00	3	5-3

FLIGHT PLANNING BRANCH

MSC FORM 1186 (SEP 67)

FLIGHT PLAN

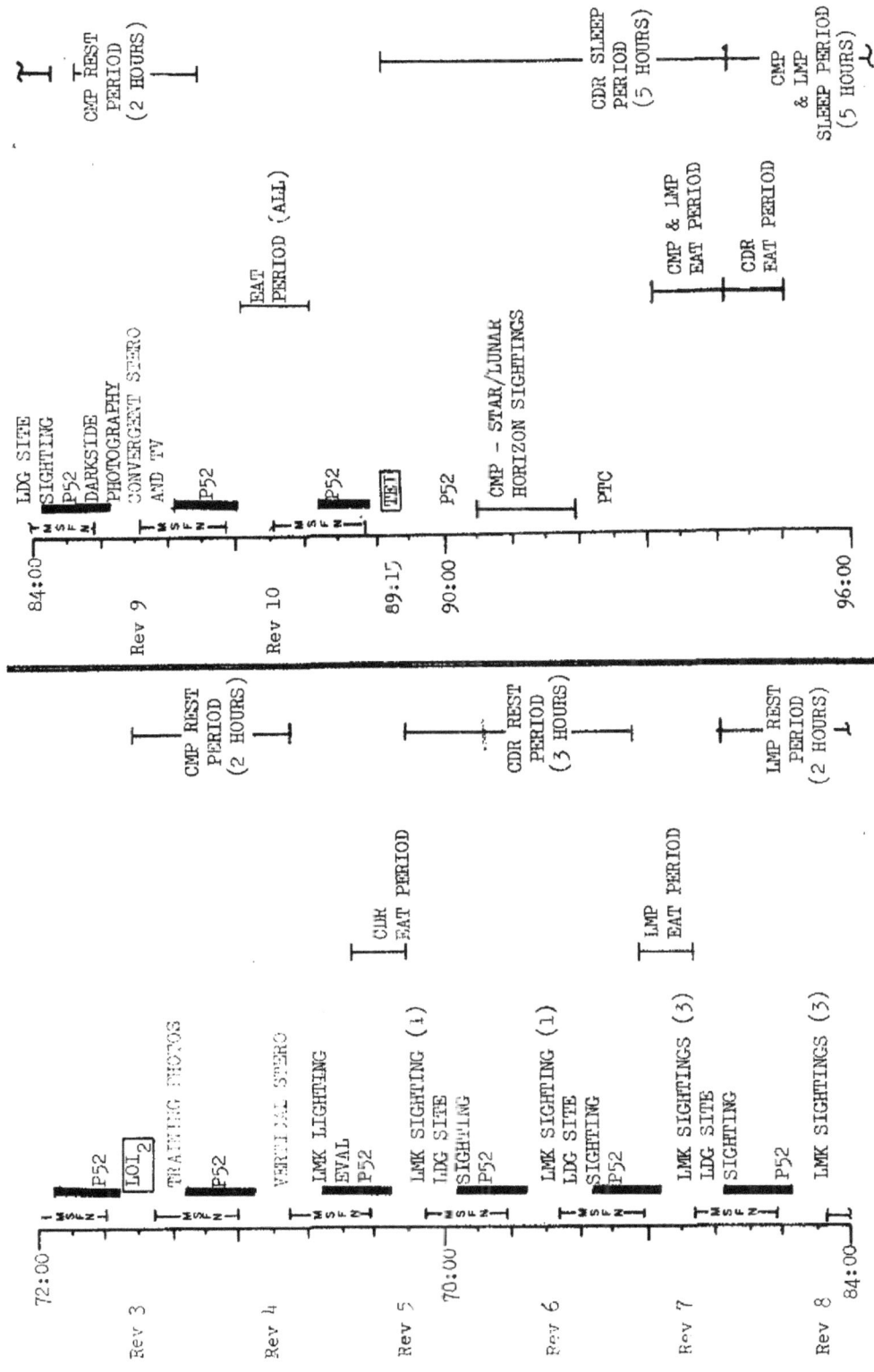

Rev 3
Rev 4
Rev 5
Rev 6
Rev 7
Rev 8

72:00
70:00
84:00

P52
LOI₂
TRAINING PHOTOS
P52
VERTICAL STERO
LMK LIGHTING EVAL
P52
LMK SIGHTING (1)
LDG SITE SIGHTING
P52
LMK SIGHTING (1)
LDG SITE SIGHTING
P52
LMK SIGHTINGS (3)
LDG SITE SIGHTING
P52
LMK SIGHTINGS (3)

CMR EAT PERIOD
LMP EAT PERIOD

CMP REST PERIOD (2 HOURS)
CDR REST PERIOD (3 HOURS)
LMP REST PERIOD (2 HOURS)

Rev 9
Rev 10

84:00
89:15
90:00
96:00

LDG SITE SIGHTING
P52
DARKSIDE PHOTOGRAPHY CONVERGENT STERO AND TV
P52
P52
TEI
P52
CMP - STAR/LUNAR HORIZON SIGHTINGS
PTC

EAT PERIOD (ALL)

CMP REST PERIOD (2 HOURS)

CMP & LMP EAT PERIOD
CDR EAT PERIOD

CDR SLEEP PERIOD (5 HOURS)
CMP & LMP SLEEP PERIOD (5 HOURS)

MISSION	EDITION	DATE	DAY/REV	PAGE
AS503/103	FINAL	November 22, 1968	4	5-4
		TIME		
		72:00 - 96:00		
		FLIGHT PLANNING BRANCH		

MSC FORM 1186 (SEP 67)

FLIGHT PLAN

96:00 — 102:00 — 108:00

P52
PTC

P52

CMP - STAR/LUNAR HORIZON SIGHTINGS

CMP - STAR/EARTH HORIZON SIGHTINGS

P52

MCC5

TEI+15

TV

CMP - STAR/EARTH HORIZON SIGHTINGS

CMP - STAR/LUNAR HORIZON SIGHTINGS

P52

EAT PERIOD ALL

CMP & LMP SLEEP PERIOD (5 HOURS)

EAT PERIOD ALL

CDR SLEEP PERIOD (7 HOURS)

108:00 — 114:00 — 120:00

CMP - STAR/EARTH HORIZON SIGHTINGS

PTC (P & Y FREE)

P52

CMP & LMP EAT PERIOD

CDR EAT PERIOD

CDR SLEEP PERIOD (7 HOURS)

CMP & LMP SLEEP PERIOD (7 HOURS)

CDR EAT PERIOD

CMP & LMP EAT PERIOD

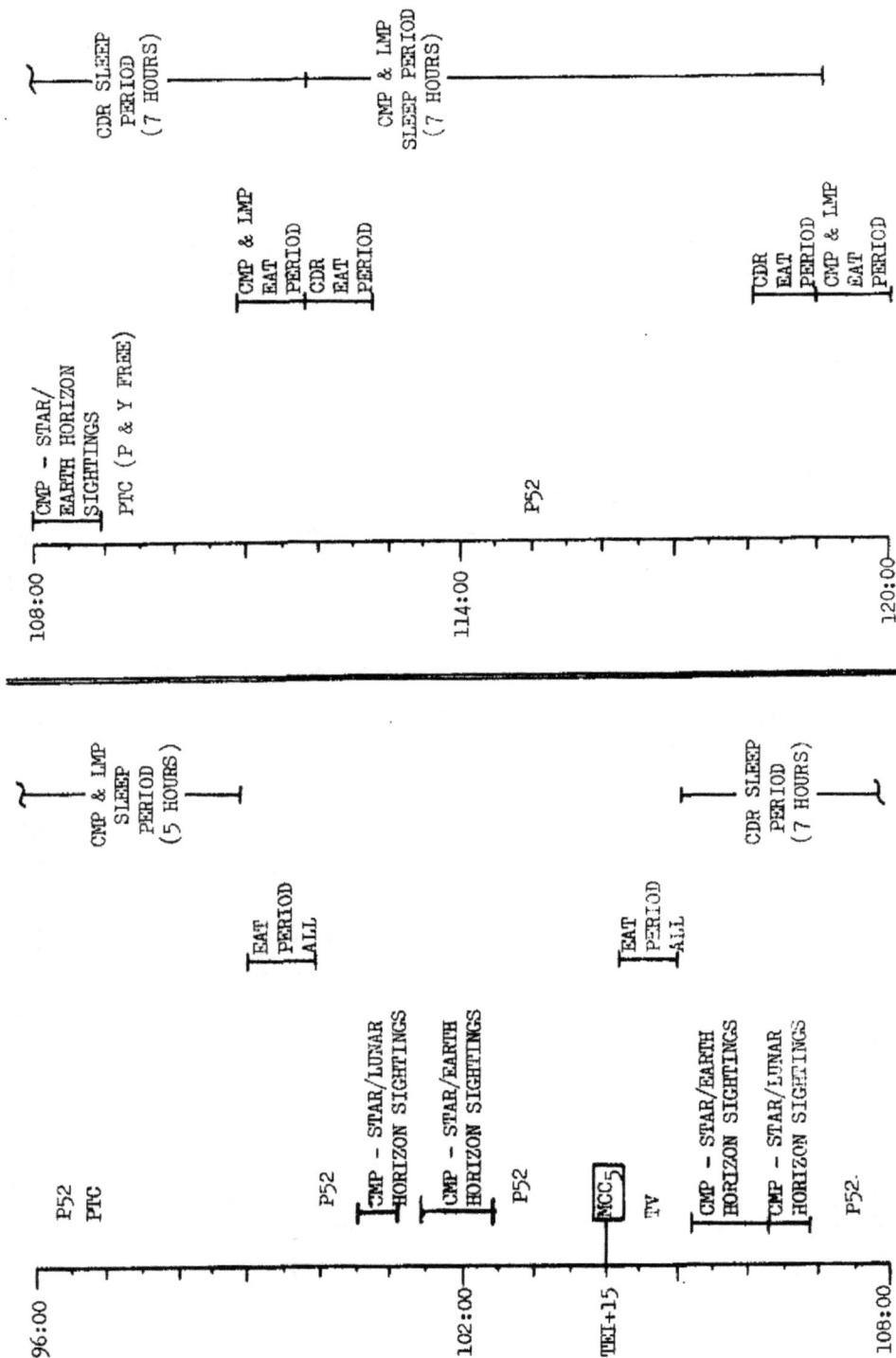

MISSION	EDITION	DATE	TIME	DAY/REV	PAGE
AS505/103	FINAL	November 22, 1968	96:00 – 120:00	5	5-5

FLIGHT PLANNING BRANCH

MSC FORM 1186 (SEP 67)

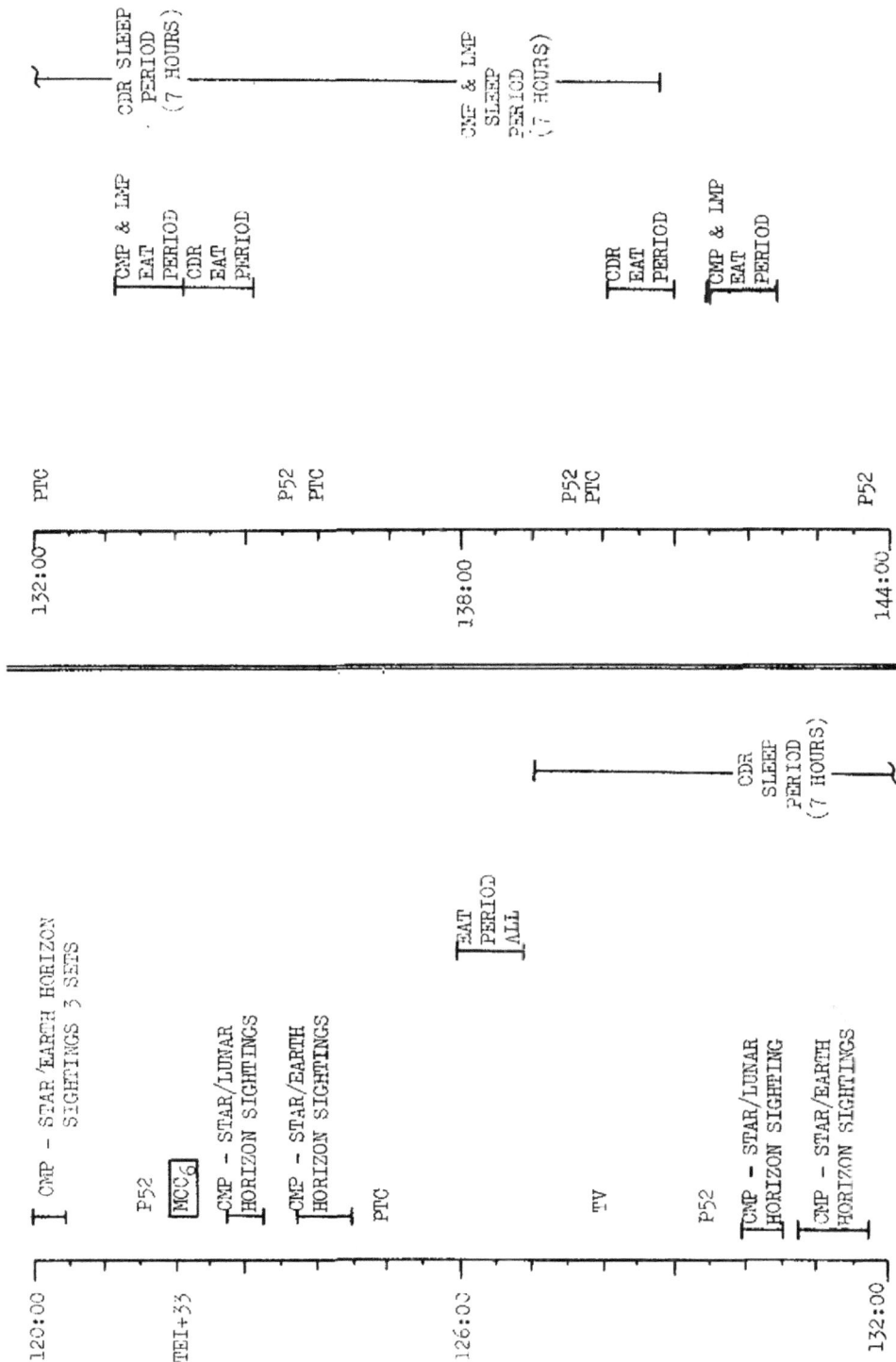

FLIGHT PLAN

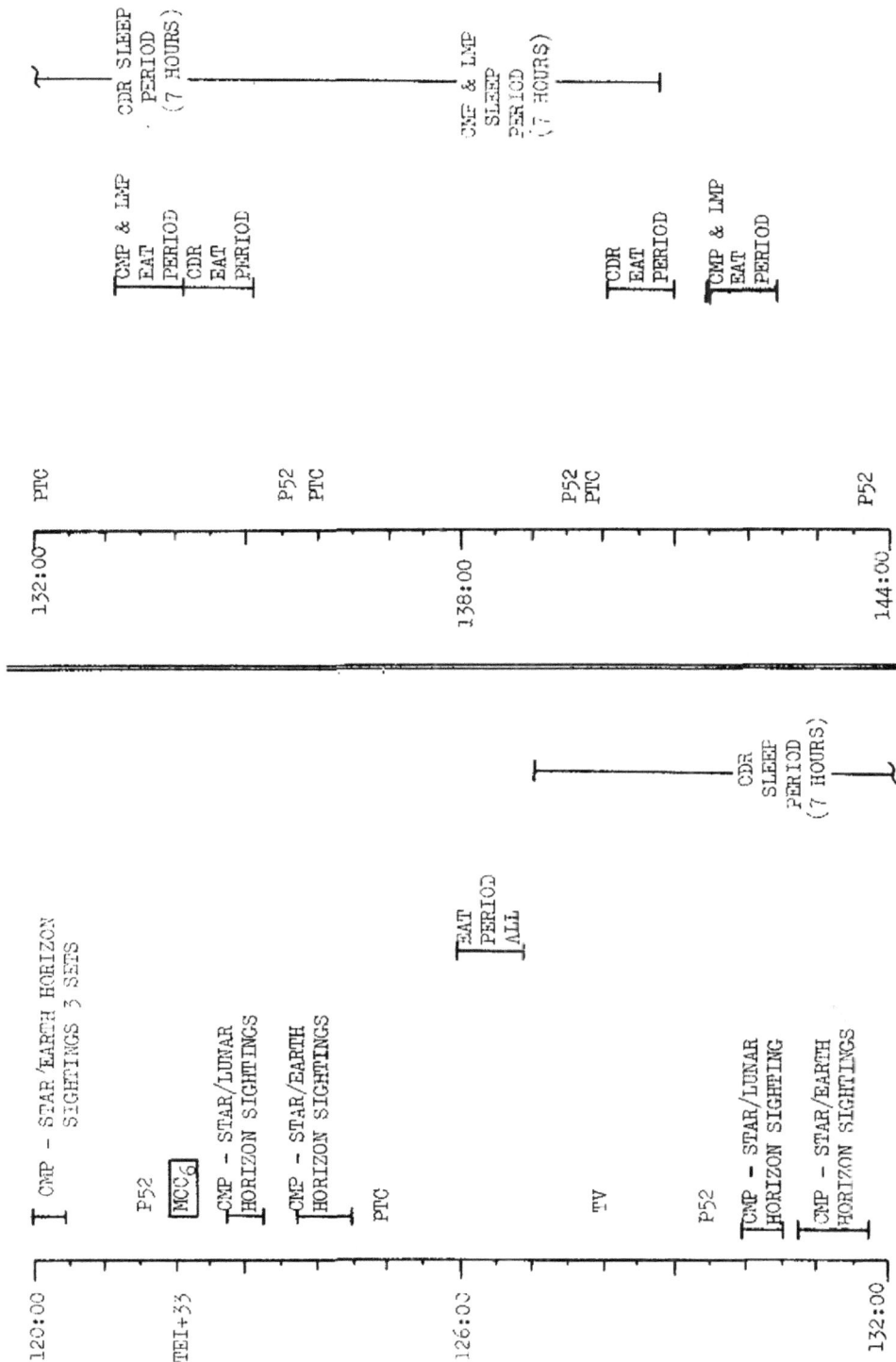

MISSION	EDITION	DATE	TIME	DAY/REV	PAGE
AS503/103	FINAL	November 22, 1968	120:00 - 144:00	6	5-6

FLIGHT PLANNING BRANCH

MSC FORM 1186 (SEP 67)

FLIGHT PLAN

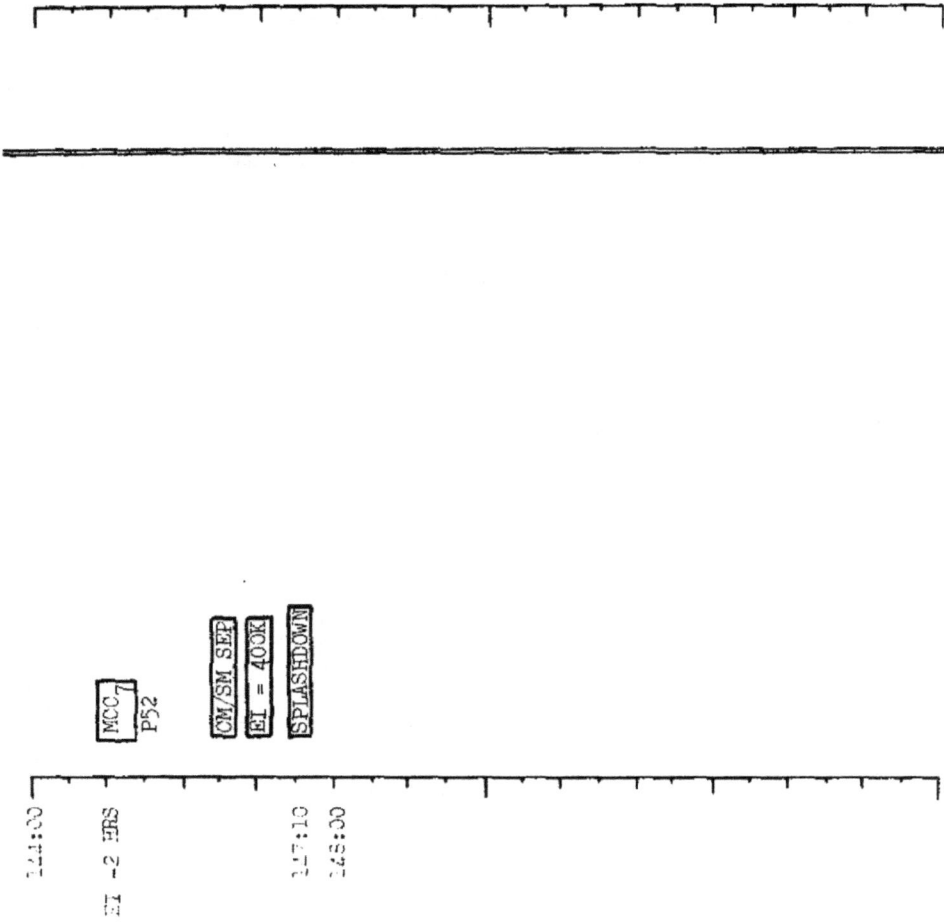

MISSION	EDITION	DATE	TIME	DAY/REV	PAGE
AS503/103	FINAL	November 22, 1968	144:00 to 146:50	6	5-7

MSC FORM 1186 (SEP 67)

FLIGHT PLANNING BRANCH

9 781780 398570